Automotive Engine Valve Recession

ENGINEERING RESEARCH SERIES

Automotive Engine Valve Recession

R Lewis and R S Dwyer-Joyce

Series Editor
Duncan Dowson

Professional Engineering Publishing Limited,
London and Bury St Edmunds, UK

First published 2002

ISBN 1 86058 358 X

ISSN 1468-3938
ERS 8

A CIP catalogue record for this book is available from the British Library.

Printed and bound in Great Britain by The Cromwell Press Limited, Wiltshire, UK.

About the Authors

Before going to university, *Dr Roger Lewis* worked for a year at the Royal Naval Engineering College in Plymouth, UK. He then studied for his MEng in Mechanical Engineering at the University of Sheffield between 1992 and 1996. During this time he was sponsored by the Ministry of Defence. Dr Lewis went on to do his PhD at the University of Sheffield (1996–1999) as part of the Tribology Research Group. His research involved the investigation of wear of diesel engine valves and seat inserts. This work was funded by the Ford Motor Company.

Dr Lewis is now a research associate at the University of Sheffield. He is currently working on railway wheel wear as part of a European project concerned with the design of a new hybrid wheel. He is also involved in a Unilever-funded project to investigate the interaction of abrasive particles and toothbrush filaments in a teeth-cleaning contact.

Professor Rob S Dwyer-Joyce is senior lecturer in the Department of Mechanical Engineering, University of Sheffield, UK. He graduated in 1993 with a PhD from Imperial College, London, where he studied the wear of rolling bearings and the effects of lubricant contamination. Before this, he worked for British Gas Exploration and Production.

Professor Dwyer-Joyce's research covers a range of tribology projects. His research group has pioneered the use of ultrasound to look at dry and lubricated engineering contacts, studied the way contaminated oil limits component life, quantified how surface damage effects railway track, and investigated aspects of automotive engine wear. He also teaches a course on the Tribology of Machine Elements to undergraduate students.

Related Titles

IMechE Engineers' Data Book – Second Edition	C Matthews	ISBN 1 86058 248 6
Design Techniques for Engine Manifolds – Wave Action Methods for IC Engines	D E Winterbone and R Pearson	ISBN 1 86058 179 X
Theory of Engine Manifold Design – Wave Action Methods for IC Engines	D E Winterbone and R Pearson	ISBN 1 86058 209 5
Statistics for Engine Optimization	Eds S P Edwards, D M Grove, and H P Wynn	ISBN 1 86058 201 X
International Journal of Engine Research		ISSN 1468/0874
Journal of Automobile Engineering	Part D of the Proceedings of the IMechE	ISSN 0954–4070

Other titles in the Engineering Research Series

Industrial Application of Environmentally Conscious Design (ERS 1)	T C McAloone	ISBN 1 86058 239 7 ISSN 1468–3938
Surface Inspection Techniques – Using the Integration of Innovative Machine Vision and Graphical Modelling Techniques (ERS 2)	M L Smith	ISBN 1 86058 292 3 ISSN 1468–3938
Laser Modification of the Wettability Characteristics of Engineering Materials (ERS 3)	J Lawrence and L Li	ISBN 1 86058 293 1 ISSN 1468–3938
Fatigue and Fracture Mechanics of Offshore Structures (ERS 4)	L S Etube	ISBN 1 86058 312 1 ISSN 1468–3938
Adaptive Neural Control of Walking Robots (ERS 5)	M J Randall	ISBN 1 86058 294 X ISSN 1468–3938
Strategies for Collective Minimalist Mobile Robots (ERS 6)	C Melhuish	ISBN 1 86058 318 0 ISSN 1468–3938
Tribological Analysis and Design of a Modern Automobile Cam Follower (ERS 7)	G Zhu and C M Taylor	ISBN 1 86058 203 6 ISSN 1468–3938

For the full range of titles published by Professional Engineering Publishing contact:

Sales Department
Professional Engineering Publishing Limited
Northgate Avenue
Bury St Edmunds
Suffolk, IP32 6BW
UK
Tel: +44 (0)1284 724384 Fax: +44 (0)1284 718692
www.pepublishing.com

Contents

Series Editor's Foreword

The nature of engineering research is such that many readers of papers in learned society journals wish to know more about the full story and background to the work reported. In some disciplines this is accommodated when the thesis or engineering report is published in monograph form – describing the research in much more complete form than is possible in journal papers. The Engineering Research Series offers this opportunity to engineers in universities and industry and will thus disseminate wider accounts of engineering research progress than are currently available. The volumes will supplement and not compete with the publication of peer-reviewed papers in journals.

Factors to be considered in the selection of items for the Series include the intrinsic quality of the volume, its comprehensive nature, the novelty of the subject, potential applications, and the relevance to the wider engineering community.

Selection of volumes for publication will be based mainly upon one of the following: single higher degree theses; a series of theses on a particular engineering topic; submissions for higher doctorates; reports to sponsors of research; or comprehensive industrial research reports. It is usual for university engineering research groups to undertake research on problems reflecting their expertise over several years. In such cases it may be appropriate to produce a comprehensive, but selective, account of the development of understanding and knowledge on the topic in a specially prepared single volume.

Volumes have already been published under the following titles:

ERS1	*Industrial Application of Environmentally Conscious Design*
ERS2	*Surface Inspection Techniques*
ERS3	*Laser Modification of the Wettability Characteristics of Engineering Materials*
ERS4	*Fatigue and Fracture Mechanics of Offshore Structures*
ERS5	*Adaptive Neural Control of Walking Robots*
ERS6	*Strategies for Collective Minimalist Mobile Robots*
ERS7	*Tribological Analysis and Design of a Modern Automobile Cam and Follower*

Authors are invited to discuss ideas for new volumes with Sheril Leich, Commissioning Editor, Books, Professional Engineering Publishing Limited, or with the Series Editor.

The present volume, which is the eighth to be published in the Series, comes from the University of Sheffield and is entitled:

Automotive Engine Valve Recession
by
Dr R. Lewis and Dr R. S. Dwyer-Joyce
The University of Sheffield

This volume follows closely the topic of the previous volume on Automobile Cams and Followers from the University of Leeds. The coincidence of successive volumes devoted to the topic of valves in automotive engines is a clear indication of current interest in these vital engineering components.

In this volume the authors outline the essential features of valve operation and the potentially serious problems associated with wear and valve recession in automobile engines since the introduction of lead-replacement and low-sulphur fuels. The authors then outline an experimental study of valve wear and the development of design tools carried out in the Department of Mechanical Engineering in the University of Sheffield.

The control of gas flow into and out of engine cylinders still presents a major challenge to the tribologist. The authors consider the fundamental nature of contact and wear between valves and valve seats and they outline the development of a special apparatus for the simulation of engine operating conditions. Valve wear and its effect upon engine performance will continue to be of concern for some time to come.

This latest volume represents a valuable addition to the Engineering Research Series. It will be of particular interest to students of wear, designers and manufacturers of reciprocating engines, valve train specialists, and tribologists.

Professor Duncan Dowson
Series Editor
Engineering Research Series

Authors' Preface

Valve wear has been a serious problem to engine designers and manufacturers for many years. Although new valve materials and production techniques are constantly being developed, these advances have been outpaced by demands for increased engine performance. The drive for reduced oil consumption and exhaust emissions, the phasing out of leaded petrol, reductions in the sulphur content of diesel fuel, and the introduction of alternative fuels such as gas all have implications for valve and seat insert wear.

This book aims to provide the reader with a complete understanding of valve recession, starting with a brief introduction to valve operation, design, and operating conditions such as loading and temperature. A detailed overview of work carried out previously, looking at valve and seat wear, is then given and valve and seat failure case studies are discussed.

A closer look is then taken at work carried out at the University of Sheffield, UK, including the development of purpose-built test apparatus capable of providing a simulation of the wear of valves and seats used in automotive engines. Experimental investigations are carried out to identify the fundamental valve and seat wear mechanisms, the effect of engine operating parameters on wear, and to rank potential new seat materials.

An important aspect of research is the industrial implementation of the results and the provision of suitable design tools. A design procedure is outlined, which encapsulates the review of literature, analysis of failed specimens, and bench test work. This includes a semi-empirical model for predicting valve recession run in an iterative software programme called RECESS, as well as flow charts to be used to reduce the likelihood of recession occurring during the design process and to offer solutions to problems that do occur.

R Lewis and R S Dwyer-Joyce
The University of Sheffield, UK

Notation

Unless otherwise stated the notation used is as follows:

A	Wear area (m^2)
b	Valve disc thickness (m)
c_i	Initial valve clearance (m)
e	Valve energy (J)
E	Modulus of elasticity (N/m^2)
f	Actuator sinusoidal displacement cycle frequency (Hz)
h	Penetration hardness (N/m^2)
k	Sliding wear coefficient
K	Empirically determined impact wear constant
l	Valve lift (m)
l_a	Actuator lift (m)
L	Initial actuator displacement (m)
m	Mass of valve + mass of follower + half mass of valve spring (kg)
n	Empirically determined impact wear constant
N	Number of cycles
p_p	Peak combustion pressure (N/m^2)
P_c	Contact force at valve/seat interface (N)
P_p	Peak combustion load (N)
r	Recession (m)
RT	Room temperature (°C)
R_d	Seat insert radius as specified in part drawing (m)
R_i	Initial seat insert radius (m)
R_v	Valve head radius (m)
s	Wear scar width (m)
t	Time (seconds)
v	Valve velocity (m/s)
v_a	Actuator velocity (m/s)
V	Wear volume (m^3)
w	Seat insert seating face width (m)
w_i	Initial seat insert seating face width (as measured) (m)
w_d	Seat insert seating face width as specified in part drawing (m)
W	Wear mass (kg)
W_v	Work done on valve during combustion in the cylinder (J)
x	Sliding distance (m)
y	Vertical deflection of valve head under combustion pressure (m)

Greek characters

α	Actuator sinusoidal displacement cycle amplitude (m)
β	Difference between valve and seat insert seating face angles (°)
δ	Slip at the valve/seat insert interface (m)
θ	Camshaft rotation (°)

θ_s	Seat insert seating face angle (°)
θ_v	Valve seating face angle (°)
ν	Poisson's ratio
ω	Camshaft rotational speed (r/min)

Chapter 1

Introduction

1.1 Valves and seats

Valves (shown in-situ in an engine in Fig. 1.1) are used to control gas flow to and from cylinders in automotive internal combustion engines. The most common type of valve used is the poppet valve (shown in Fig. 1.2 with its immediate attachments). The valve itself consists of a disc-shaped head having a stem extending from its centre at one side. The edge of the head on the side nearest the stem is accurately ground at an angle – usually 45 degrees, but sometimes 30 degrees, to form the seating face. When the valve is closed, the face is pressed in contact with a similarly ground seat. It is the contact conditions and loading at this interface that will have the largest influence on the rate at which valve and seat wear will occur, so understanding these is a key in determining the mechanisms that cause valve recession.

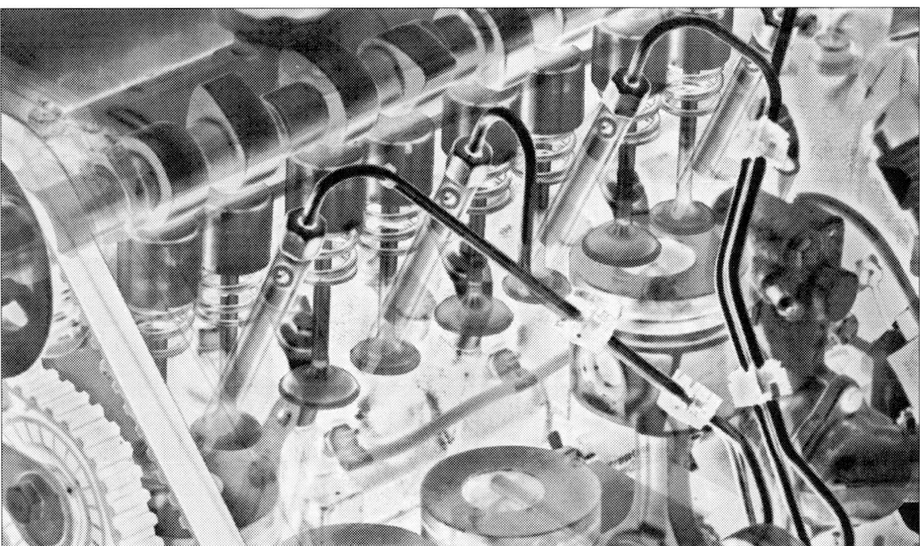

Fig. 1.1 Overhead camshaft valve drive

1.2 Valve failure concerns

Valve wear has been a small but serious problem to engine designers and manufacturers for many years. It has been described as 'One of the most perplexing wear problems in internal combustion engines' [1].

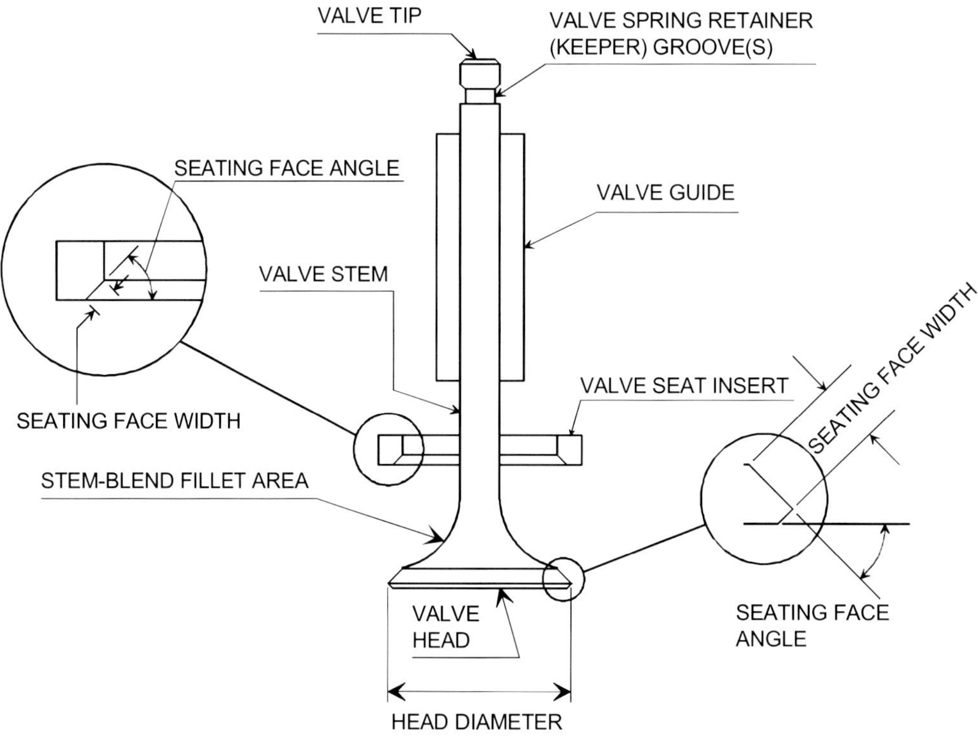

Fig. 1.2 Valve and seat insert

Although new valve materials and production techniques are constantly being developed, these advances have been outpaced by demands for increased engine performance. These demands include:

- higher horsepower-to-weight ratio;
- lower specific fuel consumption;
- environmental considerations such as emissions reduction;
- extended durability (increased time between servicing).

The drive for reduced oil consumption and exhaust emissions has led to a reduction in the amount of lubricant present in the air stream in automotive diesel engines, and the effort to lengthen service intervals has resulted in an increasingly contaminated lubricant. These changes have led to an increase in the wear of inlet valves and seat inserts.

Lead, originally added to petrol to increase the octane number, was found to form compounds during combustion that proved to be excellent lubricants, significantly reducing valve and seat wear. Leaded petrol, however, has now been phased out in the UK (since the end of 1999). As an alternative, lead replacement petrol (LRP) has been developed. This contains anti-wear additives based on alkali metals such as phosphorus, sodium, and potassium. Results of tests run using LRP containing such additives, however, have shown that, as yet, lead is unchallenged in providing the best protection. In several countries where LRP has already been introduced, a high incidence of exhaust valve burn has been recorded. In Sweden the occurrence of valve burn problems has increased by 500 per cent since LRP was introduced in 1992 [2].

The suspected cause of the valve burn problems is incomplete valve-to-valve seat sealing as a result of valve seat recession (VSR). The occurrence of VSR is blamed on hot corrosion – an accelerated attack of protective oxide films that occurs in combustion environments where low levels of alkali and/or other trace elements are present. A wide range of high-temperature alloys are susceptible to hot corrosion, including nickel- and cobalt-based alloys, which are used extensively as exhaust valve materials or as wear-resistant coatings on valves or seats. Materials used for engine components have always been designed to resist corrosive attack by lead salts. No such development has taken place to form materials resistant to alkali metals or other additive chemistries. It is clear that LRP will not provide an immediate solution to the valve wear issue, which is likely to cause tension between car manufacturers and owners for some time to come.

The impending reductions in the sulphur content of diesel fuel and the introduction of alternative fuels, such as gas, will also have implications for valve and seat insert wear. Dynamometer engine testing is often employed to investigate valve wear problems. This is expensive and time consuming, and does not necessarily help in finding the actual cause of the wear. Valve wear involves so many variables that it is impossible to confirm precise, individual quantitative evaluations of all of them during such testing. In addition, the understanding of wear mechanisms is complicated by inconsistent patterns of valve failure. For example, failure may occur in only a single valve operating in a multi-valve cylinder. Furthermore, the apparent mode of failure may vary from one valve to another in the same cylinder or between cylinders in the same engine. An example of such inconsistency is shown in Fig. 1.3. This illustrates exhaust valve recession values for four cylinders in the same engine (measurements taken on the cylinder head). The valve in cylinder 1 has recessed to the point where pressure is being lost from the cylinder, while the other valves have hardly recessed at all.

No hard and fast rules have been established to arrive at a satisfactory valve life. Each case, therefore, has to be painstakingly investigated, the cause or causes of the problem isolated, and remedial action taken. In order to analyse the wear mechanisms in detail and isolate the critical operating conditions, simulation of the valve wear process must be used. This has the added benefit of being cost effective and saving time.

Based on the wear patterns observed, the fundamental mechanisms of valve wear can be determined. Once the fundamental mechanisms are understood, a viable model of valve wear can be developed that will speed up the solution of future valve wear problems and assist in the design of new engines.

1.3 Layout of the book

Chapters 2 and 3 are review chapters outlining valve function, different operating systems, the operating environment, and valve design and materials. Valve failure is also examined in detail, and work on likely wear mechanisms and the effect of engine operating parameters are described.

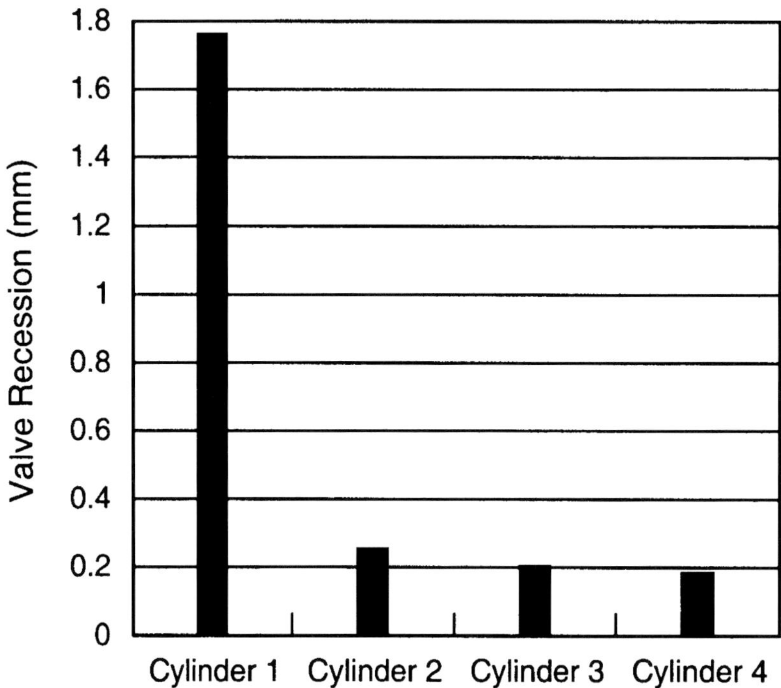

Fig. 1.3 Recession measurements for exhaust valves in the same 2.5 litre diesel engine cylinder head

Chapter 4 details the evaluation of failed valves and seat inserts from tests run on automotive engines. This includes the validation of test rig results, the establishment of techniques for the evaluation of test rig results, and the provision of information on possible causes of valve recession.

Chapter 5 outlines experimental apparatus able to simulate the loading environment and contact conditions to which the valve and seat insert are subjected in an engine.

Chapter 6 then describes bench test work carried out to investigate the wear mechanisms occurring in valves and seat inserts. This includes studies on the effect of engine operating conditions, the effect of lubrication at the valve/seat insert contact, and the evaluation of potential new seat materials.

Finally, Chapter 7 describes the development of design tools that enable the results of the review of literature, analysis of failed specimens, and bench test work to be applied in industry to assess the potential for valve recession and solve problems more quickly.

1.4 References

1. **De Wilde, E.F.** (1967) Investigation of engine exhaust valve wear, *Wear*, **10**, 231–244.

2. **Barlow, P.L.** (1999) The lead ban, lead replacement petrol and the potential for engine damage, *Indust. Lubric. Tribol.*, **51**, 128–138.

Chapter 2

Valve Operation and Design

2.1 Valve operation

2.1.1 Function

The two main types of internal combustion engine are: spark ignition (SI) engines (petrol, gasoline, or gas engines), where the fuel ignition is caused by a spark; and compression ignition (CI) engines (diesel engines), where the rise in pressure and temperature is high enough to ignite the fuel. Valves are used in these engines to control the induction and exhaust processes.

Both types of engine can be designed to operate in either two strokes of the piston or four strokes of the piston. The four-stroke operating cycle can be explained by reference to Fig. 2.1. This details the position of the piston and valves during each of the four strokes.

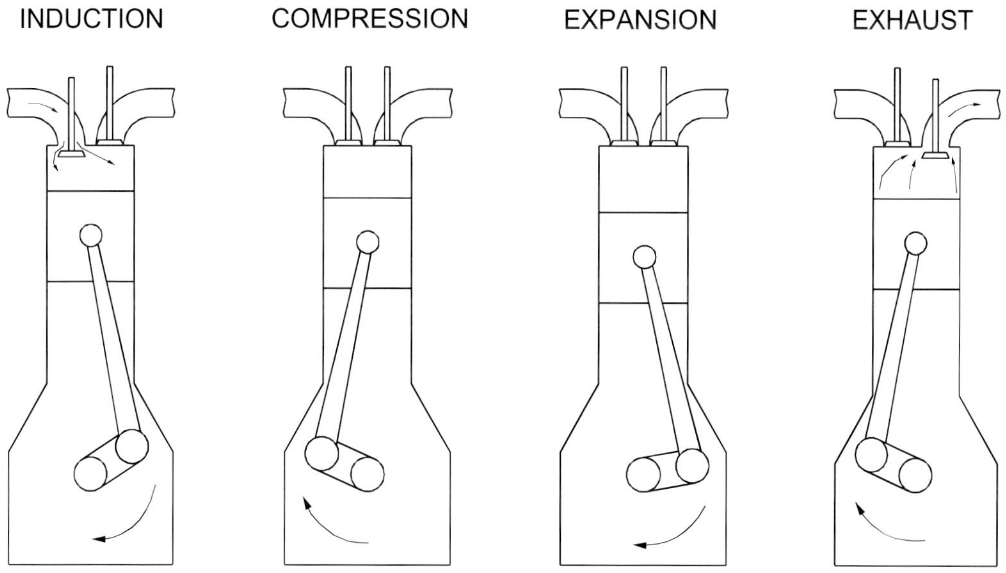

Fig. 2.1 Four-stroke engine cycle

1. **The induction stroke** The inlet valve is open. The piston moves down the cylinder drawing in a charge of air.
2. **The compression stroke** Inlet and exhaust valves are closed. The piston moves up the cylinder. As the piston approaches the top of the cylinder (top dead centre – tdc) ignition occurs. In engines utilizing direct injection (DI) the fuel is injected towards the end of the stroke.
3. **The expansion stroke** Combustion occurs causing a pressure and temperature rise which pushes the piston down. At the end of the stroke the exhaust valve opens.
4. **The exhaust stroke** The exhaust valve is still open. The piston moves up forcing exhaust gases out of the cylinder.

2.1.2 Operating systems

In engines with overhead valves (OHV), the camshaft is either mounted in the cylinder block, or in the cylinder head with an overhead camshaft (OHC).

Figure 2.2 shows an OHV drive in which the valves are driven by the camshaft via cam followers, push rods, and rocker arms. Since the drive to the camshaft is simple (either belt or chain) and the only machining is in the cylinder block, this is a cost-effective arrangement.

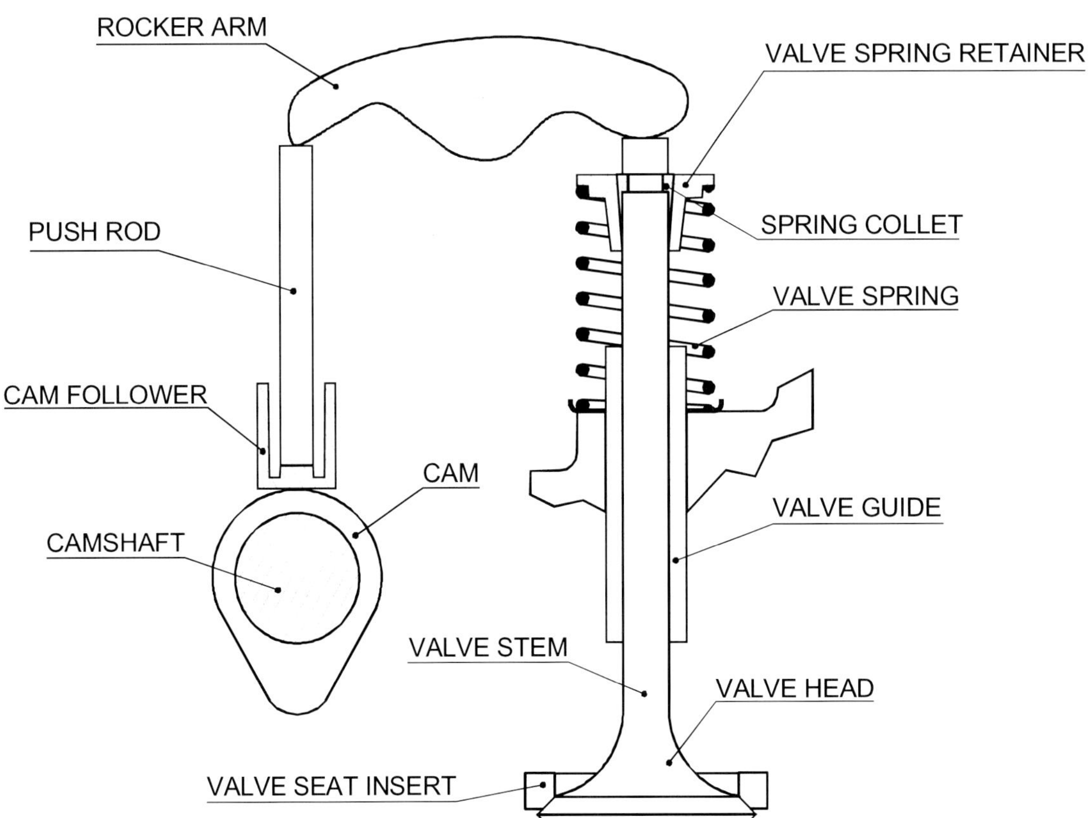

Fig. 2.2 Overhead valve drive

In the OHC drive shown in Fig. 2.3 the camshaft is mounted directly over the valve stems. Alternatively it could be offset and the valves operated using rockers. The valve clearance could then be adjusted by altering the pivot height. Once again, the drive to the camshaft is by toothed belt or chain.

In the system shown in Fig. 2.3 the camshaft operates on a follower or bucket. The clearance between the valve tip and the follower is adjusted by a shim. This is more difficult to adjust than in systems using rockers, but is less likely to change. The spring retainer is attached to the valve using a tapered split collet. The valve guides are usually press-fitted into the cylinder head, so that they can be replaced when worn. Valve seat inserts are used to ensure minimal wear. The valves rotate in order to ensure even wear and to maintain good seating. This rotation is promoted by having the cam offset from the valve stem axis. This also helps to avoid localized wear on the cam follower.

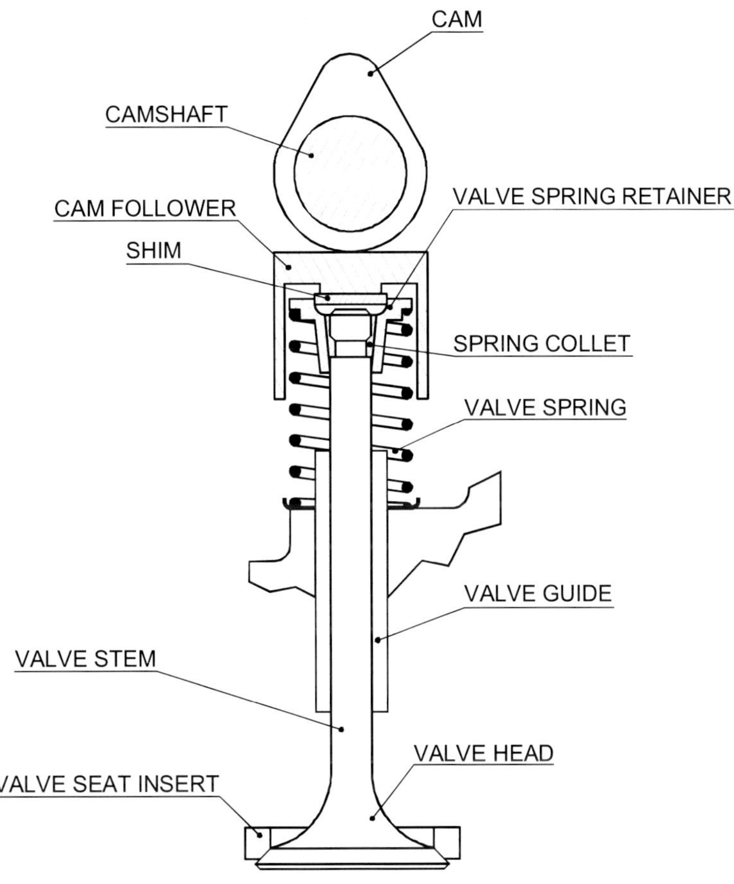

Fig. 2.3 Overhead camshaft valve drive

2.1.3 Dynamics

The geometry of the cam and its follower defines the theoretical valve motion. The actual valve motion is modified because of the finite mass and stiffness of the elements in the valve train.

Figure 2.4 [1] shows theoretical valve lift, velocity, and acceleration. The motion can be split into five stages.

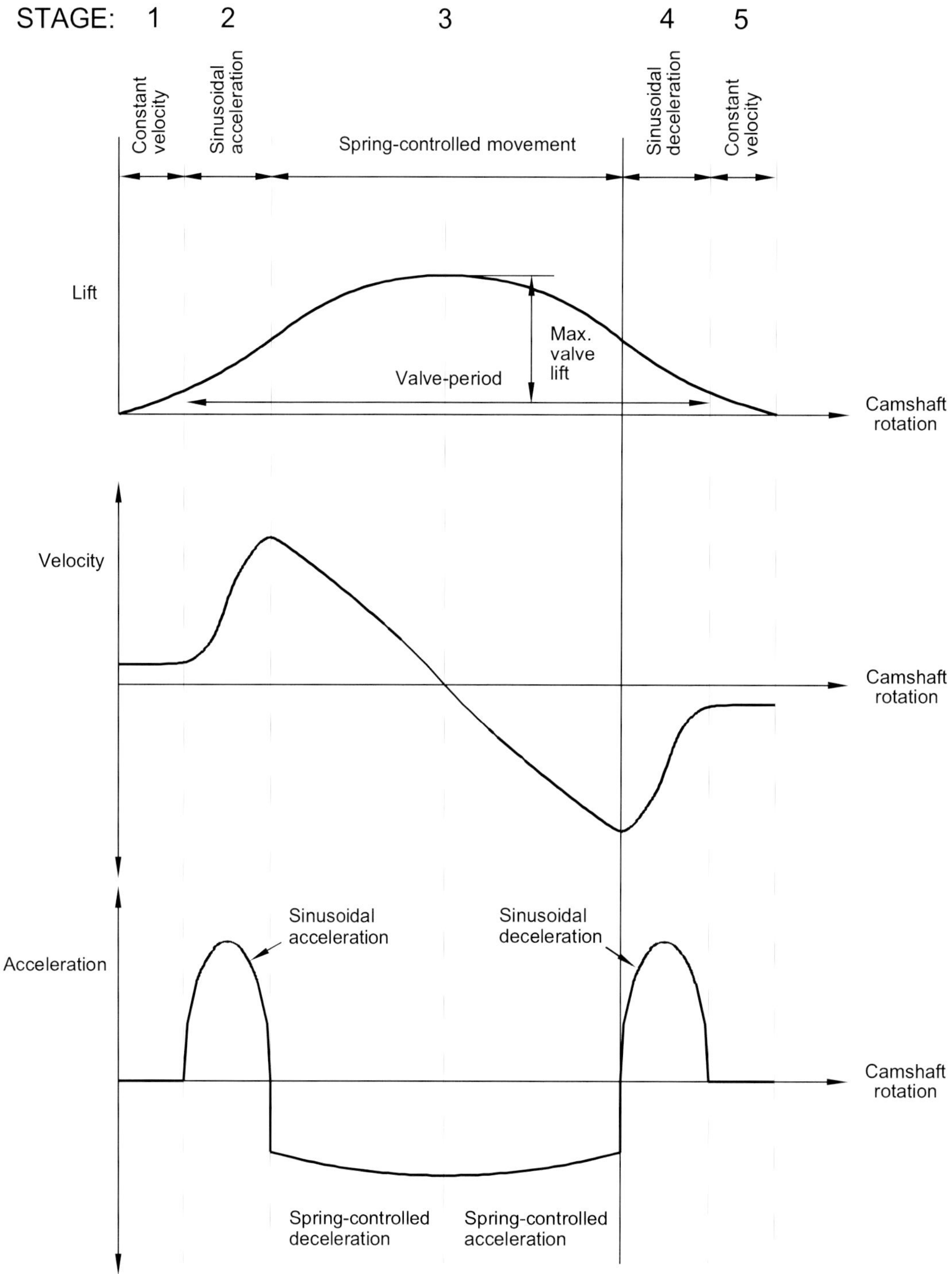

Fig. 2.4 Theoretical valve motion [1]

1. Before the valve starts to move, the clearance between the follower and the valve tip has to be taken up. This clearance ensures the valve can seat under all operating conditions and allows for bedding-in of the valve. The cam is designed to give an initially constant velocity to control the impact stresses as the clearance is taken up. The impact velocity is typically limited to 500 mm/s at the rated engine speed.
2. During the next stage the cam accelerates the valve. Rather than designing the cam to give the valve a constant acceleration (which would lead to shock loadings), sinusoidal or polynomial functions, which cause the acceleration to rise from zero to a maximum and then fall back to zero, are used.
3. Deceleration is controlled by the valve spring as the valve approaches maximum lift. As the valve starts to close, acceleration is also controlled in this way.
4. Final deceleration is controlled by the cam.
5. The cam is designed to give a constant closing velocity in order to limit impact stresses.

Actual valve motion is modified by the elasticity of the components in the valve train; a simple model is shown in Fig. 2.5.

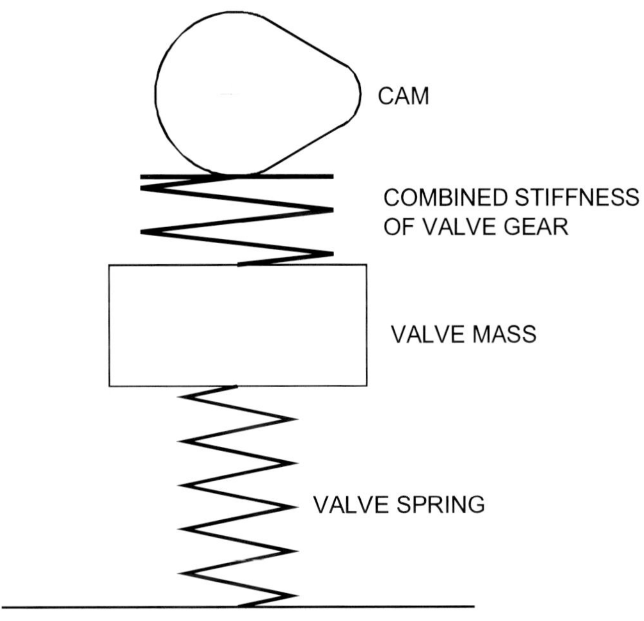

Fig. 2.5 A simple valve gear model

A comparison of actual and theoretical valve motion [1] is shown in Fig. 2.6. Valve bounce can occur if the impact velocity is too high or if the valve spring preload is too low.

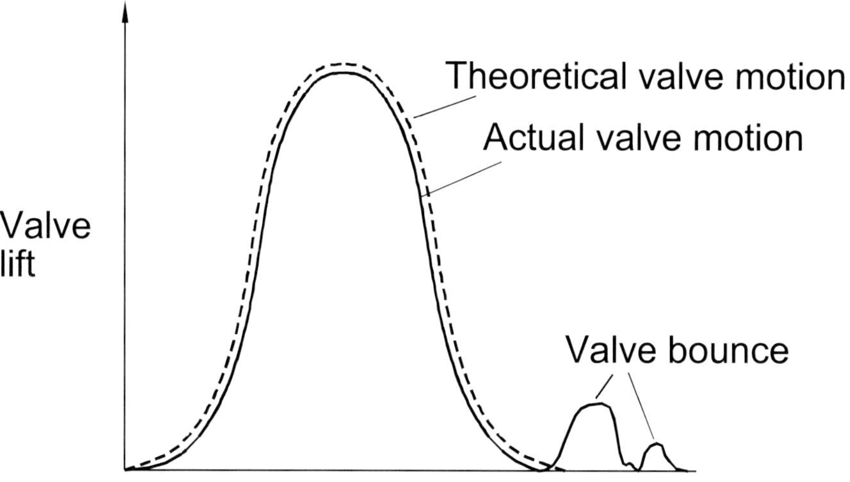

Fig. 2.6 A comparison of theoretical and actual valve motion [1]

2.1.4 Operating stresses

During each combustion event, high stresses are imposed on the combustion chamber side of the valve head. These generate cyclic stresses peaking above 200 MN/m^2 on the port side of the valve head, as shown in Fig. 2.7 [2]. The magnitude of the stresses is a function of peak combustion pressure. The stresses are much higher in a compression ignition engine than a spark ignition engine.

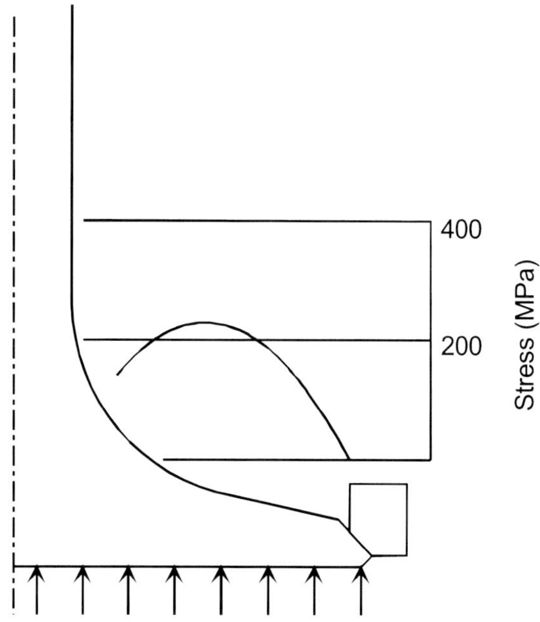

Fig. 2.7 Tensile stresses on the surface of the port side of the valve head due to combustion loading [2]

As the valve impacts the seat insert, cyclic stresses are imposed at the junction of the valve stem and fillet. If thermal distortion of the cylinder head has caused misalignment of the valve relative to the seat insert, seating will occur at a single contact point (as shown in Fig. 2.8). Bending stresses as a result of this point contact increase the magnitude of the valve seating stresses.

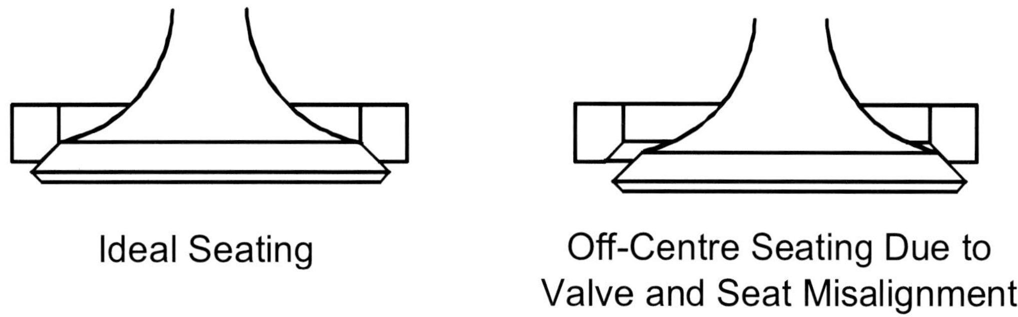

Fig. 2.8 Off-centre seating due to valve misalignment

2.1.5 Temperatures

A typical inlet valve temperature distribution is shown in Fig. 2.9 [3]. It was not made clear whether these were experimental or theoretical values or whether the valve was from a diesel or gasoline engine. The asymmetric distribution may have been due to non-uniform cooling or deposit build-up affecting heat transfer from the valve head. As shown in Fig. 2.10 [3], exhaust valve temperatures are much higher. Although both inlet and exhaust valves receive heat from combustion, the inlet valve is cooled by incoming air, whereas the exhaust valve experiences a rapid rise in temperature in the valve head, seat insert, and underhead area from hot exhaust gases.

As much as 75–80 per cent of heat input to a solid valve exits via contact with the seat insert [4]. The remainder is conducted through the valve stem into the valve guide. Effective heat transfer to the seat insert and into the cylinder head is, therefore, essential. Figure 2.11 [3] shows the thermal gradient existing from the centre of the valve head to the cooling water in the cylinder head. It clearly indicates the large temperature differential at the seating interface.

Heat transfer can be affected by valve bounce. However, the effect of seating deposits is much more significant. If deposits are allowed to build up, they may not only lead to an increase in valve temperature, but may also break away locally and create a leakage path, leading to valve guttering and, possibly, torching (see Sections 3.3 and 3.4).

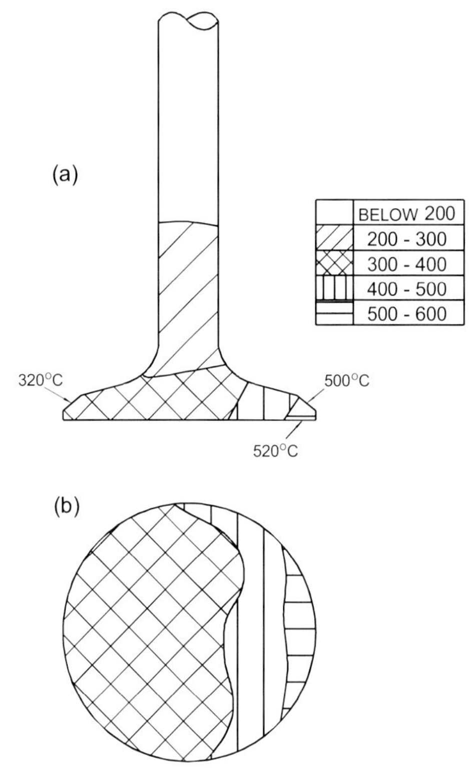

a axial section; *b* top of head

Fig. 2.9 Typical inlet valve temperature distribution [3]

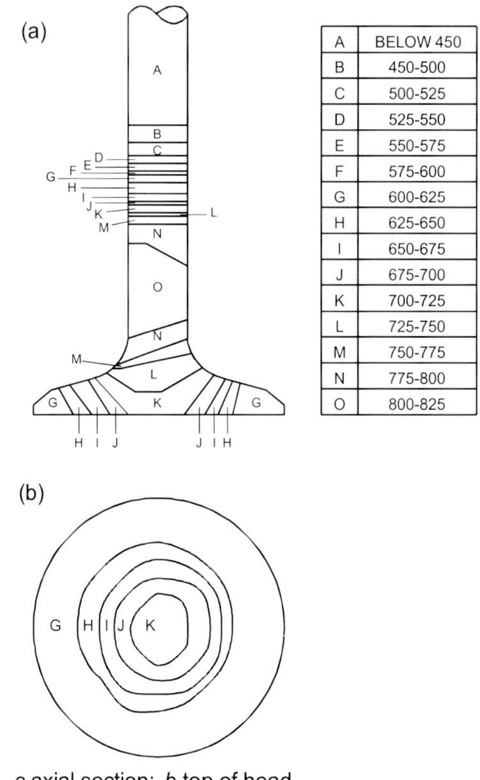

a axial section; *b* top of head

Fig. 2.10 Typical exhaust valve temperature distribution [3]

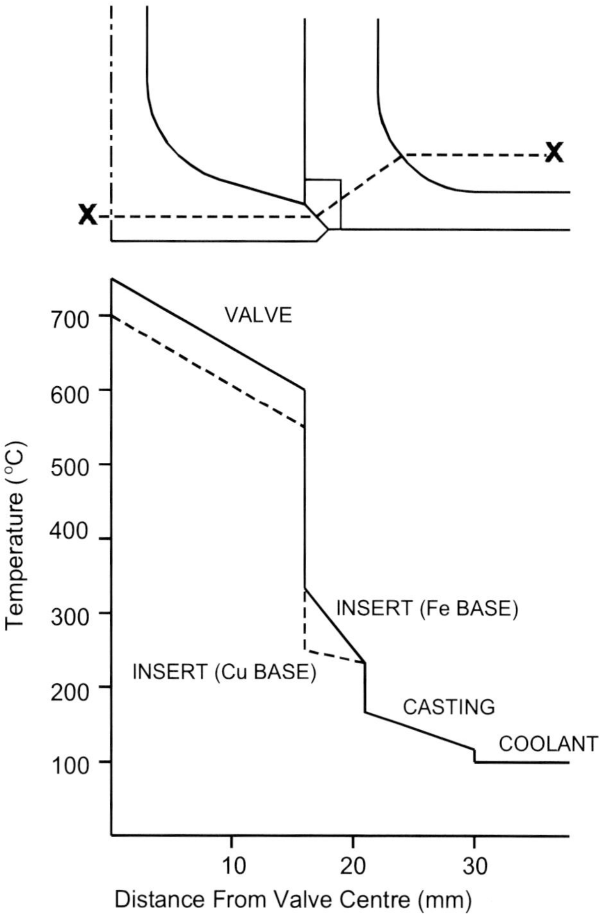

Fig. 2.11 Thermal gradient from valve centre to coolant [3]

2.2 Valve design

The most commonly used valve is the poppet valve. It has several advantages over rotary and disc valves [1]: it is cheap, has good flow properties, good seating, easy lubrication, and good heat transfer to the cylinder head.

2.2.1 Poppet valve design

A number of different poppet valve designs are used, as shown in Fig. 2.12 and outlined in Table 2.1. The final choice usually depends on the performance and cost objectives.

Inlet valves are usually constructed using the one-piece design, whereas exhaust valves are generally constructed using a two-piece design.

Valve seats formed in-situ within the cast iron cylinder heads were originally used in passenger car engines. In order to provide greater wear resistance, hardfacing alloys were developed as well as seat inserts designed to be press-fitted into the cylinder heads.

An inlet valve seat insert has been designed that is shaped to induce a swirling motion

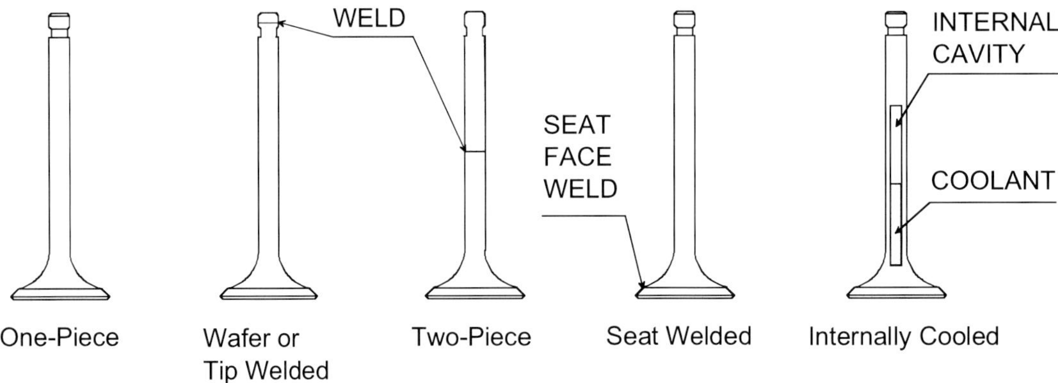

Fig. 2.12 Typical valve designs

Table 2.1 Typical valve designs

Design	Description
One-piece	This is the most cost-effective design. It is widely used in passenger car applications.
Wafer or tip welded	To eliminate tip wear or scuffing in one-piece austenitic valves, it is possible to weld on a hardened martensitic steel tip.
Two-piece	In a two-piece design an austenitic head is welded to a hardened martensitic stem. This increases both valve tip and stem scuffing resistance.
Seat welded	To increase the wear and/or corrosion resistance of the valve seating face it is possible to apply hardfacing alloys using gas or shielded-arc techniques.
Internally cooled	Internally cooled valves contain a cavity partially filled with a coolant, usually sodium, which dissipates heat from the valve head through the stem and valve guide to the cylinder head. This reduces the valve head temperature significantly.

in the fuel–air mix as it passes through the valve [5]. This swirling motion improves the mixing of the two components and enhances combustion.

The seating faces of both valves and seat inserts are usually ground to an angle of 45 degrees. The seat insert seating face width is narrower than that of the valve to reduce the risk of trapping combustion particles and wear debris in the interface between the two. In a few cases, the valve seating face is ground to an angle about half a degree less than the seating face angle, as shown in Fig. 2.13. There are three reasons for this [6].

1. The hottest part of the valve, under running conditions, is the underside of the head. The additional expansion of this side makes the two seating face angles equal at running temperatures.

2. When the valve is very hot the spring load can cause the head to dish slightly, which can lift the inner edge of the valve seating face (nearest the combustion chamber) clear of the seat insert if the angles are the same when cold.
3. The risk of trapping combustion particles between the two seating faces is reduced.

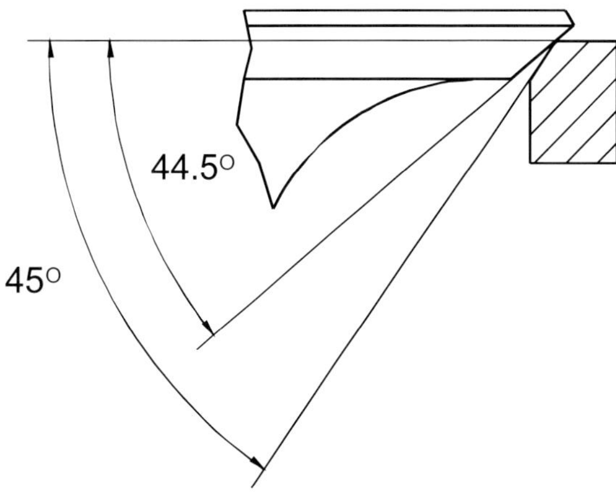

Fig. 2.13 Valve seating face angle different from seat insert seating face angle (exaggerated) [6]

2.2.2 Materials

Most inlet valves are manufactured from a hardened, martensitic, low-alloy steel. These provide good strength and wear and oxidation resistance at higher temperatures.

Exhaust valves are subjected to high temperatures, thermal stresses, and corrosive gases. Most exhaust valves are manufactured from austenitic stainless steels. These can be iron, or nickel, based. Solid solution and precipitation strengthening provide the hot hardness and creep resistance required for typical exhaust valve applications. The 21.4N composition is widely used in diesel engine exhaust valve applications. This alloy has an excellent balance of hot strength, corrosion resistance, creep resistance, fatigue resistance, and wear properties at an acceptable cost [3]. In heavy-duty diesel engine applications higher strengths and creep resistances are attained by using superalloys as valve materials. Valve seating face wear and corrosion can be reduced by applying seat facing materials. Stellite facings are commonly used for passenger car applications. Typical compositions of martensitic and austenitic steels, superalloys, and seat facing materials used in valve applications are shown in Table 2.2.

Engine test work carried out using ceramic valves has shown a significant reduction in valve recession compared to results achieved with metal valves [7, 8]. The reduction in mass of ceramic valves results in improved seating dynamics, reducing seating forces and eliminating valve bounce. In addition, the high stiffness of ceramics helps resist flexing of the valve head, reducing sliding between the valve and seat.

Table 2.2 Compositions of typical valve and seat materials [2, 3]

Nominal compositions of martensitic valve materials (weight %)

Designation	C	Mn	Si	Cr	Ni	Mo	Fe	Other	
SAE 1541	0.40	1.50	0.23	–	–	–	Bal.		
SAE 1547	0.47	1.50	0.23	–	–	–	Bal.		
SAE 3140	0.40	0.80	0.28	0.65	1.25	–	Bal.		
SAE 4140	0.40	0.88	0.28	0.95	–	0.20	Bal.		
Silchrome 1	0.45	0.80 max	3.25	8.50	0.50 max	–	Bal.		
Sil XB	0.80	0.80 max	2.12	20.00	1.35	–	Bal.	W	1.08
								V	0.25
422 SS	0.22	0.75	0.50 max	11.75	0.75	1.08	Bal.	W	18.00
								V	1.00

Nominal compositions of austenitic valve materials (weight %)

Designation	C	Mn	Si	Cr	Ni	N	Fe	Other	
21.2N	0.55	8.25	0.25 max	20.38	2.12	0.30	Bal.		
21.4N	0.53	9.00	0.25 max	21.00	3.88	0.46	Bal.		
21.12	0.20	1.50	1.00	21.25	11.50	–	Bal.		
23.8N	0.33	2.50	0.75	23.00	8.00	0.32	Bal.		
Silchrome 10	0.38	1.05	3.00	19.00	8.00	–	Bal.		
Gaman H	0.52	12.25	2.65	21.25	–	0.45	Bal.		
XCR	0.45	0.50	0.50	23.50	4.80	–	Bal.	Mo	1.08
YXCR	0.40	0.80	0.80	24.00	3.80	–	Bal.	Mo	1.40
TPA	0.45	0.60	0.60	14.00	14.00	–	Bal.	W	2.40
								Mo	0.35

Nominal compositions of superalloy valve materials (weight %)

Designation	C	Mn	Si	Cr	Ni	Fe	Others	
N-155	0.12	1.50	1.00 max	21.25	19.50	Bal.	Co	19.75
							Mo	3.50
							W	2.50
							Nb	1.00
TPM	0.04	2.25	0.08	16.0	Bal.	6.50	Ti	3.05
Inconel 751	0.06	0.50 max	0.50 max	15.50	Bal.	7.00	Ti	2.30
							Al	1.22
							Nb+Ta	0.95
Nimonic 80A	0.10 max	1.00 max	1.00 max	19.50	Bal.	3.0 max	Ti	2.25
							Al	1.40
Pyromet 31	0.04	0.20 max	0.20 max	22.60	56.50	Bal.	Ti	2.25
							Al	1.25
							Mo	2.00
							Nb	0.85

Nominal compositions of typical valve seat facing materials (weight %)

Designation	C	Mn	Si	Cr	Ni	Co	W	Mo	Fe
Stellite 6	1.20	0.50	1.20	28.00	3.00	Bal.	4.50	0.50	3.00
Stellite F	1.75	0.30	1.00	25.00	22.00	Bal.	12.00	–	3.00
Stellite 1	2.50	0.50	1.30	30.00	1.50	Bal.	13.00	0.50	3.00
Eatonite	2.40	0.50	1.00	29.00	Bal.	10.00	15.00	–	8.00
Eatonite 3	2.00	0.50	1.20	29.00	Bal.	–	–	5.50	8.00
Eatonite 6	1.75	0.75	1.30	28.00	16.50	–	–	4.50	Bal.
VMS 585	2.25	–	1.00	24.00	11.00	–	–	5.50	Bal.

Valve seats were formed in-situ within the cast iron cylinder heads in most passenger car engines. These proved inadequate in heavier-duty engines so, in order to provide greater wear resistance, hardfacing alloys were developed as well as seat inserts designed to be press-fitted into the cylinder heads. Nickel- and iron-base alloys are commonly used as hardfacing and insert materials. Sintered seat insert materials incorporating solid lubricants have also been developed to compensate for the reduction of lead and sulphur in fuels [5, 9]. These have found use in gasoline engines, but have been largely unsuccessful in diesel engines because of the higher loads. More recently, work carried out on ceramic seat insert materials has shown that they may offer a solution to valve recession problems [7, 10]. However, while ceramic materials offer unique properties that make them ideally suited for application in passenger car engines, they are relatively brittle, which raises the issue of component reliability.

2.3 References

1. **Stone, R.** (1992) *Introduction to internal combustion engines*, Macmillan, Basingstoke.

2. **Larson, J.M., Jenkins, L.F., Belmore, J.E.,** and **Narasimhan, S.L.** (1987) Engine valves – design and material evolution, *Trans ASME J. Engng Gas Turbines Power*, **109**, 351–361.

3. **Beddoes, G.N.** (1992) Valve materials and design, *Ironmaking and steelmaking*, **19**, 290–296.

4. **Giles, W.** (1971) Valve problems with lead free gasoline, SAE Paper 710368.

5. **Lane, M.S.** and **Smith, P.** (1982) Developments in sintered valve seat inserts, SAE Paper 820233.

6. **Hillier, V.A.W.** (1991) *Fundamentals of motor vehicle technology*, Fourth edition, Stanley Thornes (Publishers) Ltd, Cheltenham.

7. **Kalamasz, T.G.** and **Goth, G.** (1988) The application of ceramic materials to internal combustion engines, SAE Paper 881151.

8. **Updike, S.H.** (1989) A comparison of wear mechanics with ceramic and metal valves in firing engines, SAE Paper 890177.

9. **Fujiki, F.** and **Makoto, K.** (1992) New PM seat insert materials for high performance engines, SAE Paper 920570.

10. **Woods, M.E.** and **McNulty, W.D.** (1991) Ceramic seats and intermetallic coated valves in a natural gas fired engine, SAE Paper 910951.

Chapter 3

Valve Failure

3.1 Introduction

Three types of valve failure have been observed:

- valve recession;
- guttering;
- torching.

Valve recession, the most common form of wear in diesel engine inlet valves, is caused by loss of material from the seat insert and/or the valve. Guttering is a high-temperature, corrosive process usually caused by deposit flaking. Torching or melting of a valve is triggered by a rapid rise in the temperature of the valve head, which may be caused by preignition or abnormal combustion.

Inlet valve wear is a particular problem in diesel engines because the fuel is introduced directly into the cylinder. The inlet valve, therefore, receives no liquid on its seating face and seats under rather dry conditions.

Exhaust valve wear is less prominent than inlet valve wear as combustion products deposited on the seating faces provide lubrication. Exhaust valves are more likely to fail due to guttering or torching. Such failures are rarely seen in inlet valves.

3.2 Valve recession

Valve recession is said to have occurred if wear of the valve and seat insert contact faces has caused the valve to 'sink' or recede into the seat insert, thereby altering the closed position of the valve relative to the cylinder head (as shown in Fig. 3.1).

Engines are typically designed to tolerate a certain amount of valve recession. After this has been exceeded, the gap between the valve tip and the follower must be adjusted to ensure that the valve continues to seat correctly. If the valve is not able to seat, cylinder pressure will be lost and the hot combustion gases that leak will cause valve guttering or torching to occur, which will rapidly lead to valve failure.

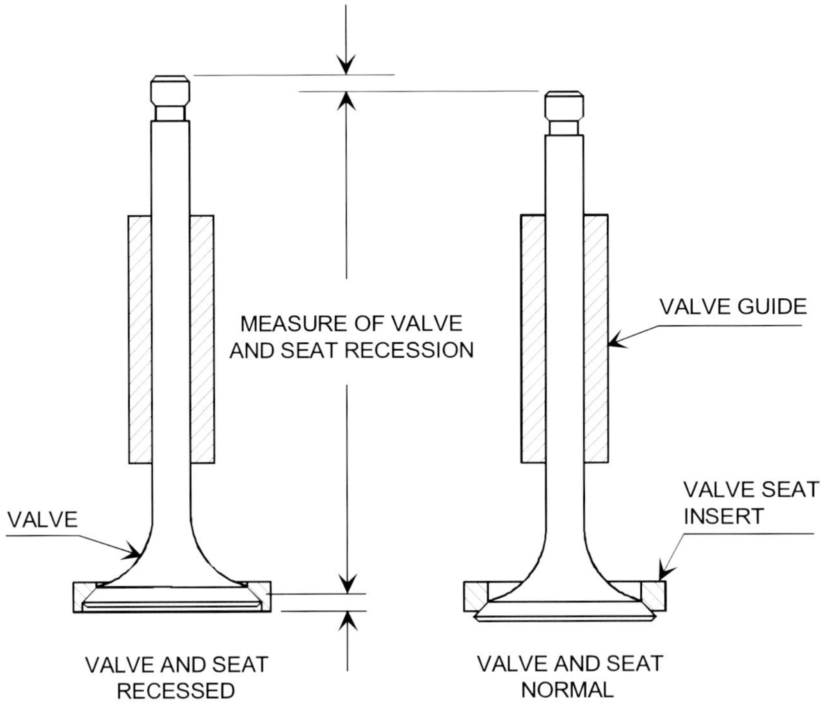

Fig. 3.1 Valve recession

3.2.1 Causes of valve recession

Valve recession is caused by loss of material from the seat insert and/or the valve. It occurs gradually over a large number of hours. Sometimes the material loss will be greater from the seat insert, and sometimes the material loss will be greater from the valve. The nature of the material loss is not clearly understood, although it has been suggested that it may occur by the following mechanisms [1]:

- metal abrasion;
- fretting;
- adhesion mechanisms;
- high temperature corrosion.

Tauschek and Newton [2] suggested that recession problems were caused by pounding of the valve due to misaligned seating, since seating contact pressure varies inversely with contact area. It was thought that improper seating was caused by cylinder head deformation due to thermal effects. Thermal effects are often associated with non-symmetric cooling passages in the cylinder head near the valve seat inserts.

Work carried out on inlet valves by Zinner [3] began on the assumption that the wear was caused by the 'hard pounding of the seat by the valve cone'. This led to an initial study of valve motion at increasing engine speeds. However, when the amount of wear under different operating conditions was measured, it became apparent that the effect of mean effective pressure on wear was much greater than that of engine speed. It was, therefore, assumed that the main factor accounting for wear must be sliding friction between the valve and valve seat caused by 'wedging' of the valve into the seat under the effect of the gas pressure. The 'wedging' action was found to be greater, and hence the sliding motion lengthened, if cylinder head deformation also occurred as a result of uneven cooling.

Figure 3.2 diagrammatically represents the effect of thermal distortions in the cylinder head. Figure 3.2(a) shows a valve not subjected to pressure on a non-deformed seat insert in a position flush with the cylinder head. The assumption in Fig. 3.2(b) is that the cylinder head has been deformed due to uneven temperatures and the valve, being under no load, projects slightly from its seat insert and makes only one-sided contact. Figure 3.2(c) shows the valve forced into its seat by the gas pressure, which implies the existence of sliding motion.

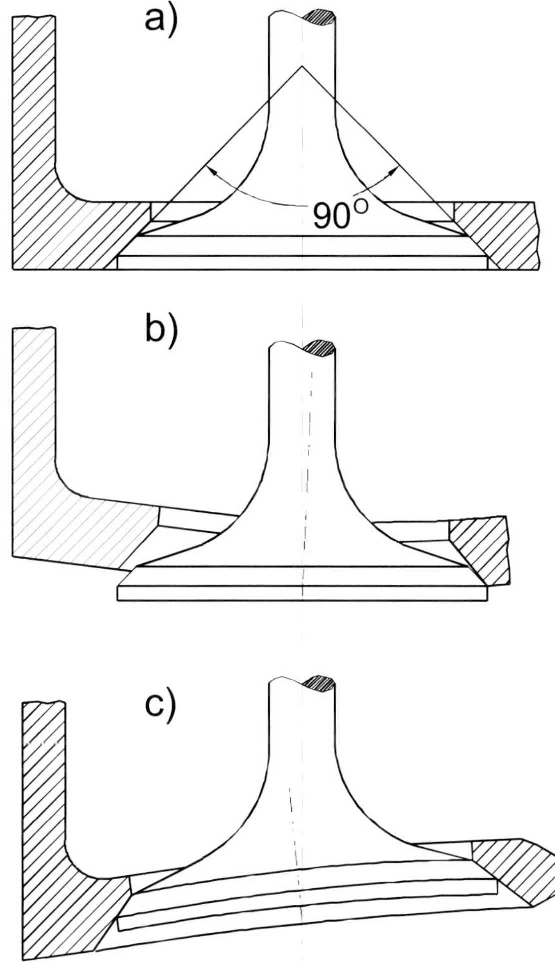

Fig. 3.2 Deformation of a cylinder head bottom and valve: (a) valve with 45 degree seat angle in plane cylinder head bottom; (b) downward bottom deflection; valve making one-sided contact; (c) bottom deflected upwards and valve disc bent under influence of gas pressure [3]

The presence of sliding motion between the valve and valve seat insert was firmly established by Zinner [3] using a static test rig designed to study the effect of seat insert distortion. Valve stem protrusion was measured at different air pressures applied to the valve head. An increase in valve stem movement resulted in greater sliding at the valve/seat contact area.

Work carried out by Marx and Muller [4] on wear of inlet valves in supercharged four-stroke diesel engines started on the supposition that wear was provoked by incorrect closing of the inlet valves (valve bounce). However, it was eventually concluded that wear was caused by small frictional movements between the valve seating face and the seat insert. It was thought that the friction movements were a result of elastic bending of the valve and working face of the cylinder head due to the combustion pressure. It was found that wear was more frequent in higher power engines due to higher engine velocity and, therefore, an increased number of combustion cycles.

Lane and Smith [5] studied the force mechanism between a valve and a valve seat insert (as shown in Fig. 3.3). Force *P*, applied to the seat insert, consists of valve train inertia force due to acceleration when the valve is seating and forces applied by the valve spring and cylinder pressure when the valve is resting on the seat. This force can be resolved into two components – $P\sin\theta$ parallel to the seating face and $P\cos\theta$ perpendicular to the seating face.

It was suggested that if the $P\sin\theta$ component exceeds the shear stress of the seat insert material, plastic shear deformation of the surface material may be induced. This could lead to crack formation and eventually, after repeated loading, to particles of material (asperities) breaking away from the surface. The wear debris may either be blown off by gas flow when the valve opens or remain on the seating surface and:

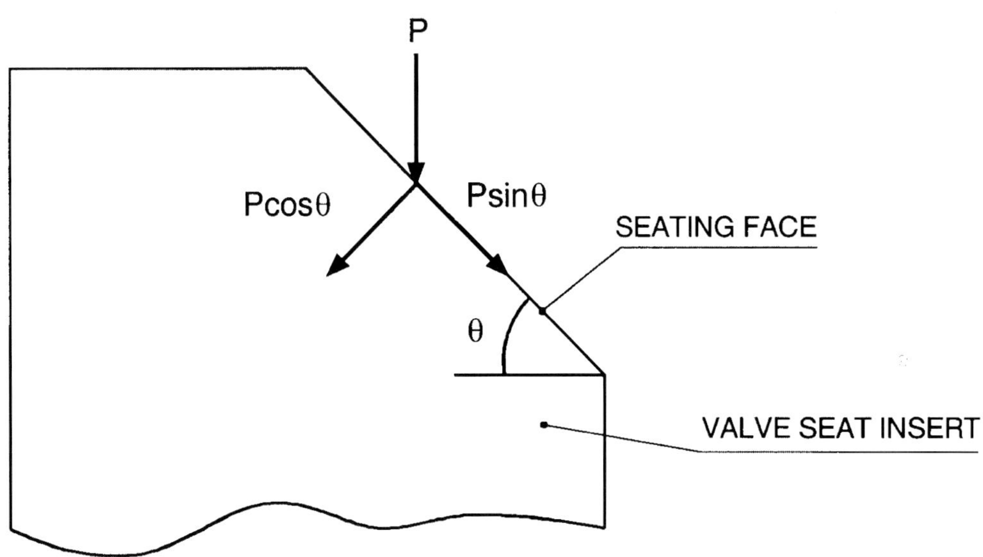

Fig. 3.3 Forces acting on a seat insert seating face [5]

1. prevent complete gas sealing;
2. lead to an increase in valve temperature due to reduced seat contact area (the reduced seat contact area inhibits heat transfer from the valve to the engine coolant through the seat insert, causing possible valve failure);
3. lead to abrasive wear of seating surfaces when the valve slides on the seat insert (if the asperities are hard material).

Using a reduced angle for the seating face is often suggested for an approach to reduce surface flow. It was calculated that a reduction in the seating angle from 45 to 30 degrees would reduce the shear force ($P\sin\theta$) by 29 per cent. The normal force ($P\cos\theta$), however, would increase by 22 per cent. It was suggested that, if the $P\cos\theta$ component exceeds the compressive yield strength of the material, it may fail in a fashion known as 'hammering' or 'brinelling'. Therefore, to compensate for a reduction in the seating angle, material hardness and toughness may need to be increased to withstand increased dynamic loading.

Work performed by Pope [6] led to the conclusion that wear of the seating face of air inlet valves was due to relative movement between the valve and its seat when the head flexed under the action of the firing pressure since:

1. the measured valve head stress due to the cylinder firing pressure was about 20 times that due to the valve impact on its seat;
2. a considerable reduction in the valve seating velocity had a negligible effect on seat wear;
3. stiffer valve gear to reduce the possibility of resonant valve gear vibrations produced little effect.

Van Dissel *et al.* [7] concluded that seat recession can occur by the systematic gouging away, or deformation and eventual wearing, of the valve and/or insert material. It was found that the deformation led to the formation of concentric ridges on the seating face of the valve which were described as 'single wave' or 'multiple wave' formations. It was suggested that the gouging and deformation were generated by the same process, only the severity of the damage was different.

It was speculated that the problem was caused by valve misalignment, which resulted in the valve seating face making contact with only a portion of the seat insert seating face. As it was expected that the initial contact would be the most severe, it is at this stage that the surface ridges were thought to be generated. Subsequent ridge formation was thought to occur as the valve bounced to self-centre, driven by the combustion cycles. Valve rotation, resulting in a different portion of the valve seating face impacting the insert for each cycle, was thought to cause the 'single' or 'multiple wave' appearance.

Both Fricke and Allen [8] and Ootani *et al.* [9], in testing materials for poppet valve applications, assumed that impact of the valve on the seat was the major cause of valve recession.

3.2.2 Wear characterization

Marx and Muller [4] found that after a relatively short running time it was possible to observe concentric rings on the seating faces of inlet valves. Local manifestations also occurred which were reminiscent of sand dunes. These changed into smooth, bright, slightly curved surfaces over time, indicating consistent removal of material. Wear rates of approximately 1 mm/1000 h were observed.

Van Dissel *et al.* [7] observed distinct trenches on the seating faces of recessed valves where seating occurred. Seat recession varied from negligible to 0.56 mm. Worn regions typically exhibited unique topographic features. Pitting, gouging, and indentation of the seating face were all observed. One feature common to most recessed valves was observed. This was a series of ridges and valleys formed circumferentially around the axis of the seating face. In some cases the ridges and valleys were concentric, as waves would appear if a single stone was dropped into a pool. This was described as a 'single wave formation' (see Fig. 3.4). In other cases the ridges and valleys overlapped each other, as waves would appear if several stones were dropped into a pool. This was described as a 'multiple wave formation' (see Fig. 3.5). Ridge formation was found to vary according to engine type.

Fig. 3.4 Inlet valve seating face showing 'single wave ridge formation' (Major diameter is towards the bottom) [7]

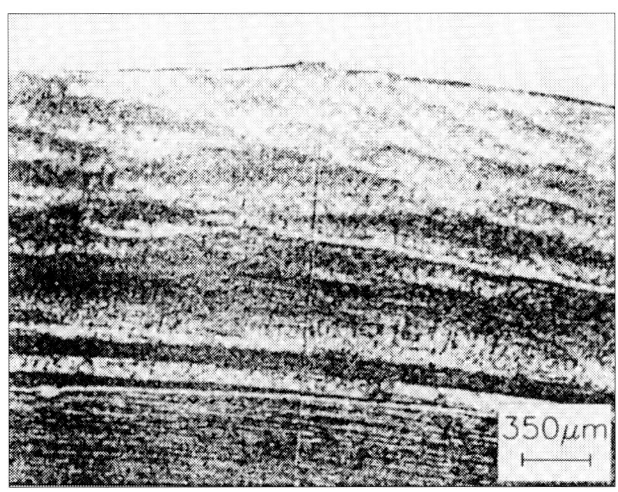

Fig. 3.5 Inlet valve seating face showing 'multiple wave formation' (Major diameter is towards the top) [7]

Wear of seat inserts ranged from negligible to moderate. The material removal process resulted in a concave seating face configuration. Radial scratches were observed within the concave region, as shown in Fig. 3.6.

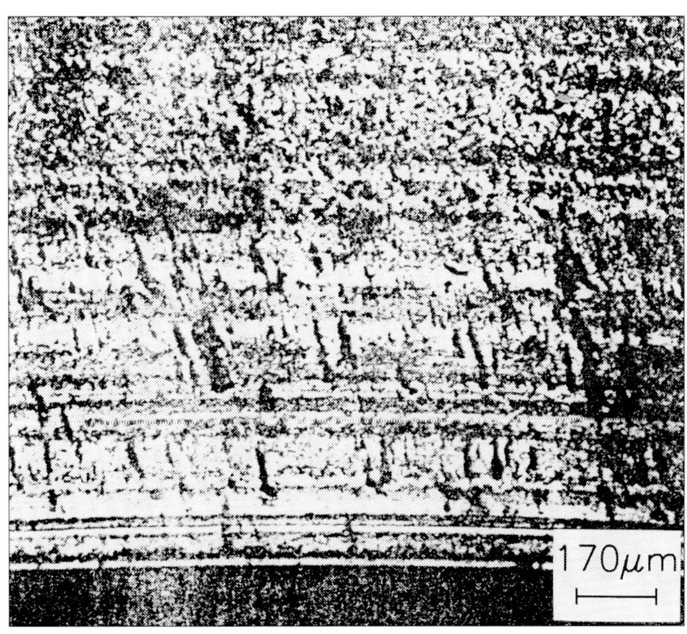

Fig. 3.6 Pitting of inlet seat insert (Major diameter is towards the top) [7]

Narasimhan and Larson [10] observed reorientation of carbides near the surface of a seat insert seating face, as shown in Fig. 3.7. This provided evidence of sliding contact between the valve and seat insert.

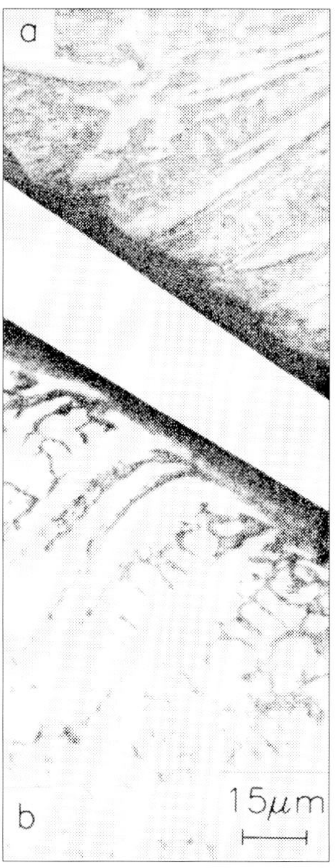

Fig. 3.7 Microstructural features of alloys at the valve/seat insert interface: (a) eatonite insert; (b) alloy No. 6 hardfacing [10]

3.2.3 Reduction of recession

Zinner [3] concluded that the major cause of valve recession was sliding friction between the valve and seat insert caused by 'wedging' of the valve into the seat under the action of the combustion pressure. In order to reduce the effect of the sliding motion, design modifications were introduced. Below is a summary of the design modifications that resulted in a reduction of wear, presented in order of effectiveness:

1. decreasing the valve seating face angle;
2. lubrication of the valve seating face;
3. use of valve seat inserts of a suitable material;
4. increased rigidity of valve disc;
5. greater rigidity of valve mechanism;

6. welding material onto valve seats;
7. smaller valve diameter;
8. reducing speed of valve impact;
9. increasing seating face width.

Introduction of the 30 degrees seating face angle, more effective cooling of the cylinder head bottom, and greater rigidity of the valve and valve mechanism reduced wear rates by about 50 per cent. It should be noted, however, as mentioned in the previous section, that changes such as reducing the valve seating face angle may not always be possible as they will change flow characteristics and may adversely affect engine performance.

In order to reduce the frictional motion, and thereby wear, Marx and Muller [4] reduced the seating face angle from 45 to 30 degrees. They then carried out investigations with over eighty material pairs. They concluded that suitable material pairing and reducing the seat angle was not enough to solve the problem. Tests were then carried out in which a small amount of lubricating oil was introduced shortly before the inlet valve. This was found to be very effective in reducing wear if used in conjunction with the changes previously detailed.

Giles [11] used an expression for predicting the depth of material lost through adhesive wear to identify parameters that should be modified in order to reduce valve recession. Tests indicated that the use of lower seat angles and hardened seat inserts showed the greatest promise in reducing valve recession.

Table 3.1 provides a summary of design modifications that have been tested in an effort to reduce valve recession.

3.3 Guttering

Guttering is a high-temperature corrosive process that usually occurs in exhaust valves. Guttering causes a leakage path to form radially across the sealing area between the valve seating face and the seat insert. In some instances the channel enlarges and, as combustion occurs in the cylinder, it follows the leak path into the valve port, rapidly melting, or 'torching' the valve [12].

Under magnification, guttered valve surfaces suffering from intergranular corrosion have a characteristic cobblestone appearance [13]. Guttering valve wear eventually causes excessive leakage of cylinder compression, and misfiring results in loss of power. A common cause for exhaust valve guttering in diesel engines is ash deposit flaking [14]. An initiating leakage path can form across a valve seating face/seat insert interface where a flake is missing. The path then becomes wider and wider.

3.4 Torching

Torching, or melting, of a valve has been observed to occur rapidly in just a few engine cycles. It is associated with engines experiencing preignition or abnormal combustion.

Table 3.1 Design changes tested in an effort to reduce valve recession

Design change	*Effect in reducing wear*
Reduction in seating face angle (45–30 degrees)	Reduced wear. With all other factors unchanged, wear was lowered by 1/3 to 1/4 of that with 45 degree angle [3]. Tests carried out with 80+ material pairs and reduction in seat angle. Concluded that change in seat angle and suitable material pairing alone would not solve problem [4]. Reduced wear [6]. 'Recession was reduced by approximately 75 per cent when seat angle was reduced from 45 to 30 degrees' [11].
Lubrication of contact between valve and seat	Reduced wear [3]. 'Very effective method for reducing wear' [4]. 'Considerable' reduction in seat wear obtained by injecting lubricating oil into the inlet manifold [6].
Reducing speed of valve impact	Reduced wear [3]. 'The most powerful means of reducing wear is to keep impact velocities as low as possible' [8]. 'A considerable reduction in valve seating velocity had negligible effect on seat wear' [6].
Use of hardened valve seat inserts	Reduced wear [6]. 'Recession rates were reduced 80–95 per cent by using hardened inserts' [11].
Positive rotation of valve	Reduced wear [3].
Greater rigidity of valve head	Reduced wear [3]. 'Stiffer valve head sections were found to be beneficial in reducing wear rates. Lower valve head deflection or 'oil canning' reduces scrubbing distance' [11].
Improved cooling of cylinder head bottom	Reduced wear [3].
Greater rigidity of valve mechanism	Reduced wear [3].
Welding material onto valve seating face	Reduced wear [3].
Increasing seat width	Reduced wear [3]. 'Increased seat width showed little reduction in wear rates' [11].
Hardening of seat	Flame hardening of seats had little effect in reducing wear [3]. Induction hardening of seats reduced wear rate 25 per cent over that with non-hardened seats [11].
Reducing clearance between stem and valve guide	Little effect in reducing wear [3].
Reducing valve head weight	Little effect in reducing wear [3].
Using slightly different valve and seat angles	Little effect in reducing wear [3].
Use of resilient valve head (tulip valve)	Increased wear [3].
Use of resilient seat insert	Increased wear [3].

Preignition is described as unscheduled premature combustion. Some part of the combustion chamber becomes hot enough to ignite the fuel–air charge prior to timed ignition from the spark plug. Once initiated, the cycle continues with the combustion chamber becoming hotter as ignition occurs earlier and earlier in the cycle. If unchecked the severe temperature rise literally melts the more susceptible components in the combustion chamber, usually the piston crown or the exhaust valve. Preignition failures of this type are seldom seen in diesel/gas engines because there is no fuel mixed with the air during early compression.

With abnormal combustion ('knock') there is a high pressure rise before or near piston top-dead-centre. The high pressure rise causes compression heating of burning gases and a more rapid heat release rate, raising the valve temperature which may trigger valve torching.

3.5 Effect of engine operating parameters

3.5.1 Temperature

Inlet valve temperatures are not normally high enough to cause significant corrosion or thermal fatigue failures. Such failures are far more likely to occur in exhaust valves. However, a recent study of inlet valve failures [15] led to the conclusion that deposit build-up on the seating face of an inlet valve (formed from engine oil and fuel) had reduced heat transfer from the valve head (a valve transfers approximately 75 per cent of the heat input to the top-of-head through its seat insert into the cylinder head [11]), resulting in tempering and reduced hardness. As a result, some valves had suffered failures due to radial cracking of the seating face induced by thermal fatigue while others had failed due to valve guttering.

Cherrie [16] found that when the temperature of a 21.4N steel valve was increased from 704 to 732 °C, the stress that could be sustained to rupture (failure accompanied by significant plastic deformation) in 100 hours decreased by 35 per cent. De Wilde [17] found, however, that at the temperatures experienced in exhaust valves and seat inserts, there was no significant reduction in mechanical properties and thus discounted temperature as a major influence on valve wear.

Matsushima [18] investigated the wear rate of valve and seat inserts at elevated temperatures. Several insert materials were tested with Stellite valves. Figures 3.8 and 3.9 show the wear rate of the exhaust valve and seat insert, respectively. Valve seating face wear increased as the temperature exceeded 200 °C and continued to rise as the temperature was increased to 500 °C. At temperatures below 200 °C the wear was almost negligible. The wear rate of the seat inserts peaked at 300 °C, decreased at 400 °C, but rose again as the temperature was increased above 400 °C. The use of superalloy seat inserts reduced the seat insert wear below that of cast iron inserts, but increased the wear on the valve seating face. The powder metal insert containing materials to form lubricious oxides improved both seat insert and valve wear.

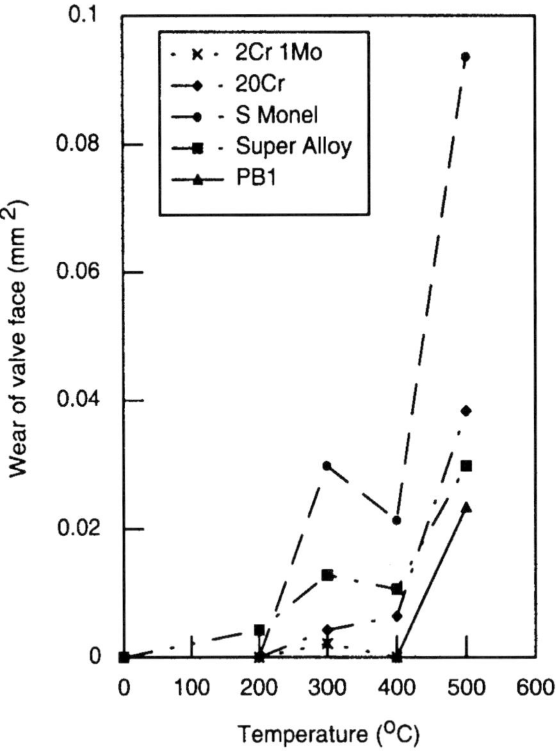

Fig. 3.8 Wear of Stellite valve faces when mated with various seat inserts [18]

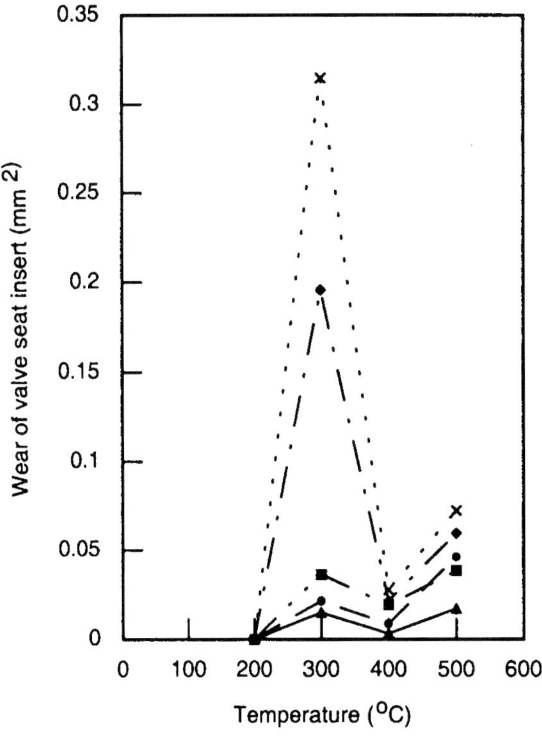

Fig. 3.9 Wear of various seat inserts when mated with Stellite valves [18]

Hofmann *et al.* [19] and Wang *et al.* [20], however, have shown that the general trend is that wear decreases as temperature increases (as shown in Fig. 3.10). This was thought to be because of oxide formation at high temperatures preventing metal-to-metal contact and thus reducing adhesive wear.

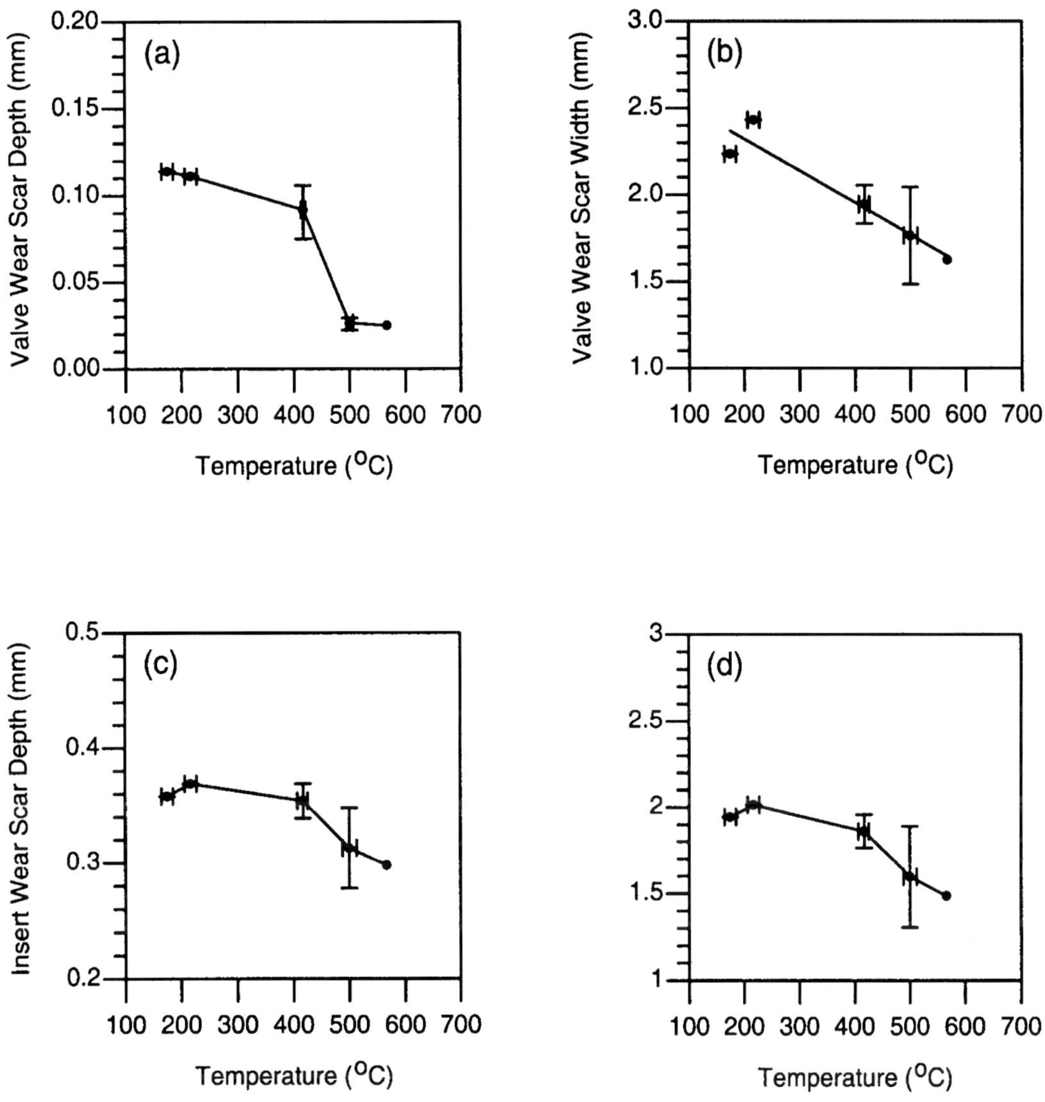

Fig. 3.10 Seating face wear as a function of temperature: (a) valve seating face scar depth; (b) valve seating face scar width; (c) seat insert seating face scar depth; (d) seat insert seating face scar width [20]

3.5.2 Lubrication

Engine lubricants have been found to both increase and reduce valve and seat wear, depending on the additive composition and the amount of oil that reaches the valve/seat interface.

In inlet valves, liquid film lubrication is most dominant as temperatures are not usually high enough to volatilize the lubricant hydrocarbons and additives. Exhaust valves, however, are predominantly lubricated by solid films formed at the higher operating temperatures by oil additive ash compounds such as alkaline-earth and other metal oxides, sulphates, and phosphates (e.g. calcium, barium, magnesium, sodium, zinc, and molybdenum). Very thin metal oxide films have been found to be beneficial in reducing valve wear [21]. Too much solid film lubricant, however, can be detrimental and lead to valve guttering or torching due to flaking (as described in Sections 3.3 and 3.4).

3.5.3 Deposits

Extensive work has been carried out investigating the formation mechanisms, effect, and methods of reducing inlet valve deposits in gasoline engines [22–28]. This has shown that deposits are produced from engine oil, fuel, and soot-like particles [24, 27] and that deposits accumulating on inlet valves affect drivability, exhaust emissions, and fuel consumption in gasoline engines. Many experiments have demonstrated that engine parameters, such as oil leakage at valve guides and positive crankshaft ventilation, valve temperature, and exhaust gas recirculation influence the deposit formation. Very little work, however, covers inlet valve deposits in diesel engines.

It has been shown that exhaust valve deposits, formed from combustion products, prove favourable in providing lubrication on the seat contact surface (see Section 3.5.2). Their influence on inlet valve wear, however, has not been investigated.

Esaki *et al.* [24] characterized the deposit formation on inlet valves in diesel engines. Figure 3.11 shows this formation, as well as the temperatures of the various parts of the valve. The black deposit was made up mainly of concentrated engine oil, oxidation products, and precarbonization products. The major part of the grey deposit was ash. It mainly comprised calcium sulphate. It was demonstrated that the black deposits accumulate at inlet valve temperatures of approximately 230 to 300 °C. At inlet valve temperatures above 350 °C the deposits or components of engine oil on the inlet valve were converted to ash.

3.5.4 Rotation

Valve rotation can be achieved either by the use of positive rotators or by the use of multi-groove collets rather than clamping collets. These allow the valve to rotate under vibrational influences from the valve train or valve spring. This rotation can be promoted if the centre of the cam is offset from the valve axis.

Hiruma and Furuhama [29] measured exhaust valve rotation and found that, in the engine under consideration, the valve started rotating after the engine speed exceeded 3000 r/min and then increased rapidly at higher engine speeds (as shown in Fig. 3.12).

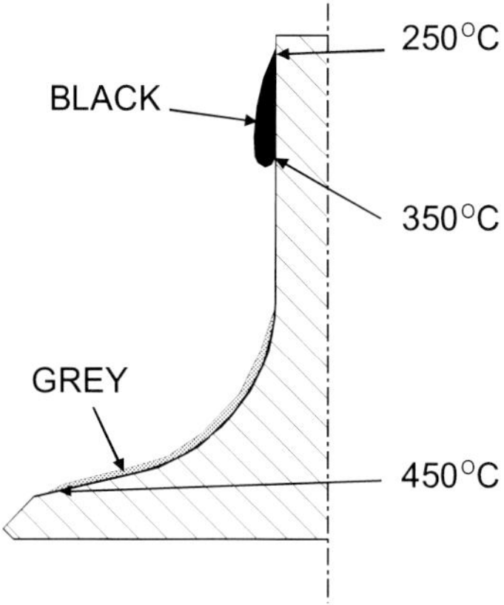

Fig. 3.11 Cross-section of accumulated deposits on diesel engine inlet valves as characterized by Esaki *et al.* [24]

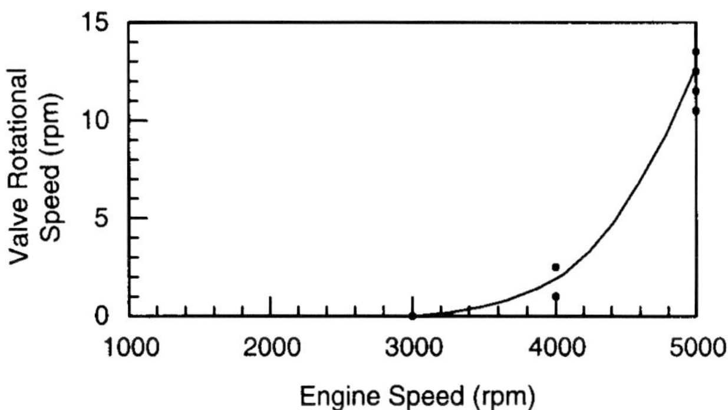

Fig. 3.12 Speed of rotation of an exhaust valve [29]

It was also found that at low speed operation (3000 r/min) the valve did not rotate constantly and changed its direction occasionally.

Beddoes [30] also observed that valve rotation was random and occurred in either direction. It was found that valves stopped and started rotating as engine speed altered, usually beginning rotation at about 50 per cent of maximum engine speed.

There is some agreement that valve rotation is beneficial in grinding away deposits. This prevents local hot spots forming and helps maintain good sealing and thermal contact of the valve to the seat [30–32]. The role valve rotation plays in valve/seat insert wear, however, is not fully understood.

3.6 Summary

The review of literature indicated that the majority of the work carried out previously on diesel engine valve wear had focussed on large engines rather than those utilized in passenger cars, and a greater emphasis had been placed on investigating exhaust valve wear than that found in inlet valves. The complex nature of the valve operating environment and the difficulties associated with making a quantitative analysis of the effect of the many variables involved in the valve operating system was highlighted in the work. Most investigations had concluded that valve and seat wear was caused by frictional sliding between the valve and seat under the action of the combustion pressure. Little account was taken of other possible wear mechanisms.

Parameter studies had mainly focussed on the effect of engine operating conditions such as temperature and load. Little or no work had been carried out to investigate the effect on wear of design parameters, material properties, valve closing velocity, and the effect of reducing lubrication at the valve seat/interface.

3.7 References

1. **Pyle, W.** and **Smrcka, N.** (1993) Effect of lubricating oil additives on valve recession in stationary gaseous-fuelled four-cycle engines, SAE Paper 932780.
2. **Tauschek, M.J.** and **Newton, J.A.** (1953) Valve seat distortion, SAE Preprint 64.
3. **Zinner, K.** (1963) Investigations concerning wear of inlet valve seats in diesel engines, ASME Paper 63-OGP-1.
4. **Marx, W.** and **Muller, R.** (1968) Ein βeitrag zum Einlaβventilsitz-Verscheiβ an aufgeladeren Viertakt-Dieselmotoren (A contribution on the subject of the wear of inlet valve seats in supercharged four-stroke diesel engines – its origins and some remedies), MTZ Paper No. 29, in German.
5. **Lane, M.S.** and **Smith, P.** (1982) Developments in sintered valve seat inserts, SAE Paper 820233.
6. **Pope, J.** (1967) Techniques used in achieving a high specific airflow for high-output medium-speed diesel engines, *Trans ASME J. Engng Power*, **89**, 265–275.
7. **Van Dissel, R.**, **Barber, G.C.**, **Larson, J.M.**, and **Narasimhan, S.L.** (1989) Engine valve seat and insert wear, SAE Paper 892146.
8. **Fricke, R.W.** and **Allen, C.** (1993) Repetitive impact-wear of steels, *Wear*, **163**, 837–847.
9. **Ootani, T.**, **Yahata, N.**, **Fujiki, A.**, and **Ehia, A.** (1995) Impact wear characteristics of engine valve and valve seat insert materials at high temperature (Impact wear tests of austenitic heat-resistant steel SUH36 against Fe-base sintered alloy using plane specimens), *Wear*, **188**, 175–184.
10. **Narasimhan, S.L.** and **Larson, J.M.** (1985) Valve gear wear and materials, SAE Paper 851497, *SAE Trans*, **94**.

11. **Giles, W.** (1971) Valve problems with lead free gasoline, SAE Paper 710368.

12. **Arnold, E.B., Bara, M.,** and **Zang, D.** (1988) Development and application of a cycle for evaluating factors contributing to diesel engine valve guttering, SAE Paper 880669.

13. **McGeehan, J.A., Gilmore, J.T.,** and **Thompson, R.M.** (1988) How sulphated ash in oils causes catastrophic diesel exhaust valve failures, SAE Paper 881584.

14. **Tantet, J.A.** and **Brown, P.I.** (1965) Series 3 oils and their suitability for wider applications, NPRA, Tech 65-29L.

15. **Pazienza, L.** (1996) 1.8 IDI chromo 193 intake valve failures, EATON Report No. 86/96.

16. **Cherrie, J.M.** (1965) Factors influencing valve temperatures in passenger car engines, SAE Paper 650484.

17. **De Wilde, E.F.** (1967) Investigation of engine exhaust valve wear, *Wear*, **10**, 231–244.

18. **Matsushima, N.** (1987) Powder metal seat inserts, *Nainen Kikan*, **26**, 52–57, in Japanese.

19. **Hofmann, C.M., Jones, D.R.,** and **Neumann, W.** (1986) High temperature wear properties of seat insert alloys, SAE Paper 860150, *SAE Trans.*, **95**.

20. **Wang, Y.S., Narasimhan, S., Larson, J.M., Larson, J.E.,** and **Barber, G.C.** (1996) The effect of operating conditions on heavy duty engine valve seat wear, *Wear*, **201**, 15–25.

21. **Wiles, H.M.** (1965) Gas engines valve and seat wear, SAE Paper 650393.

22. **Bitting, B., Gschwendtner, F., Kohlhepp, W., Kothe, M., Testroet, C.J.,** and **Ziwica, K.H.** (1987) Intake valve deposits – fuel detergency requirements revisited, SAE Paper 872117 (SP-725), *SAE Trans.*, **96**.

23. **Cheng, S.** (1992) The physical parameters that influence deposit formation on an intake valve, SAE Paper 922257.

24. **Esaki, Y., Ishiguro, T., Susuki, N.,** and **Nakada, M.** (1990) Mechanism of intake valve deposit formation: Part 1 – Characterization of deposits, SAE Paper 900151, *SAE Trans.*, **99**.

25. **Gething, J.A.** (1987) Performance robbing aspects of intake valve and port deposits, SAE Paper 872116.

26. **Houser, K.R.** and **Crosby, T.A.** (1992) The impact of intake valve deposits on exhaust emissions, SAE Paper 922259.

27. **Lepperhoff, G., Schommers, J., Weber, O.,** and **Leonhardt, H.** (1987) Mechanism of the deposit formation at inlet valves, SAE Paper 872115 (SP-725).

28. **Nomura, Y., Ohsawa, K., Ishiguro, T.,** and **Nakada, M.** (1990) Mechanism of intake valve deposit formation: Part 2 – Simulation tests, SAE Paper 900152, *SAE Trans*, **99**.

29. **Hiruma, M.** and **Furuhama, S.** (1978) A study on valve recession caused by non-leaded gasoline – measurement by means of R.I., *Bullet. JSME*, **21**, 147–160.

30. **Beddoes, G.N.** (1992) Valve materials and design, *Ironmaking and steelmaking*, **19**, 290–296.

31. **Heywood, J.B.** (1988) *Internal combustion engine fundamentals*, McGraw-Hill, London.

32. **Stone, R.** (1992) *Introduction to internal combustion engines*, Macmillan, Basingstoke.

Chapter 4

Analysis of Failed Components

4.1 Introduction

This chapter details work carried out to evaluate valves and seat inserts from durability dyno tests run on a 1.8 litre, IDI, automotive diesel engine. It then goes on to report the findings of two investigations carried out concerning the failure of valves and seat inserts. The first relates to lacquer formation on valve seating faces during durability tests on a turbocharged diesel engine. The second investigates failures during durability testing of a 1.8 litre, DI, diesel engine caused by poor valve seating due to uneven seat insert wear.

The work was undertaken in order to provide a means of comparison and validation for future bench test work and to establish analysis techniques for use during such work. It was also intended to provide information on possible causes of valve recession, and engine operating conditions and design features that may influence the wear process.

4.2 Valve and seat insert evaluation

Recessed valves and seat inserts from durability dyno tests run on a 1.8 litre, IDI, automotive diesel engine were evaluated in order to provide a comparison and validation for later bench testing. Techniques such as profilometry and optical microscopy were used for the analysis.

4.2.1 Specimen details

Valves and inserts made from a variety of materials were examined. Seat insert materials S1 and S2 consist of a sintered martensitic tool steel matrix with evenly distributed intra-granular spheroidal alloy carbides. Metallic sulphides are distributed throughout at original particle boundaries. The interconnected porosity in both is substantially filled with copper alloy throughout. Seat insert material S3 is cast and consists of a tempered martensitic tool steel matrix with a network of carbides uniformly distributed. Valve material V1 is a martensitic low-alloy steel and material V2 is an austenitic stainless steel. Some of these materials are currently used in inlet valve applications while the others were being tested to assess their performance.

4.2.2 Profile traces

Profile traces of both valve and seat insert seating faces were taken using a profilometer (Surfcom) (see Fig. 4.1). The valve and seat insert stands shown were clamped in position and the stylus was returned to the same position for each profile taken. Therefore, apart from inaccuracies due to slight differences in the machining of the valves and seat inserts, each profile had the same origin.

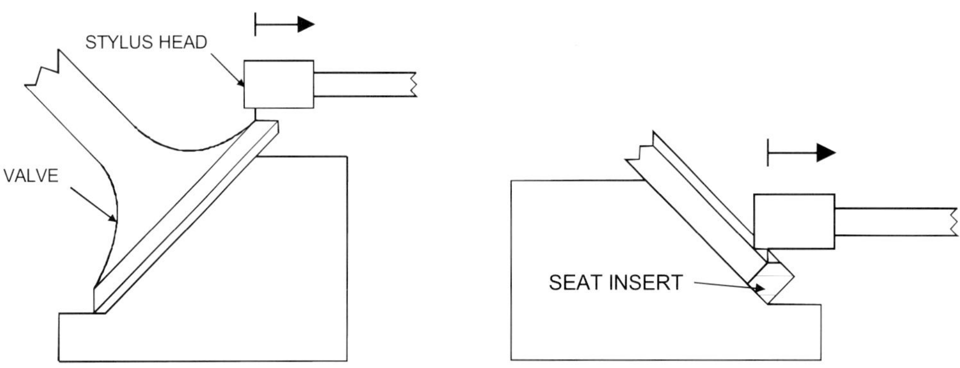

Fig. 4.1 Use of a profilometer to take profile traces of valves and seat inserts

Figures 4.2 and 4.3 show profiles of inlet valves and seat inserts taken from two dyno tests run under the same operating conditions. Comparison of the valve profiles with that of an unused valve, also shown in Fig. 4.2, gives an indication of the amount of wear that has occurred.

Valve clearance data recorded during the tests (shown in Fig. 4.4) indicated that in the first test (valve material V1 run against seat insert material S1) major valve recession occurred (0.4 mm in 100 hours against a benchmark of 0.6 mm in 250 hours), whereas in the second (valve material V1 run against seat insert material S3) only minor valve recession occurred (0.15 mm in 130 hours). This is confirmed by comparison of the seat insert profiles (see Fig. 4.3), which clearly shows that the sintered seat insert material S1 has worn more than the cast seat insert material S3. The valve wear, however, was greater in the second test. This can be explained by looking at the seat insert material used in each test. The cast seat insert material used in the second test is tougher and more resistant to impact than the sintered material used in the first test, hence the reduction in seat insert wear and the increase in valve wear.

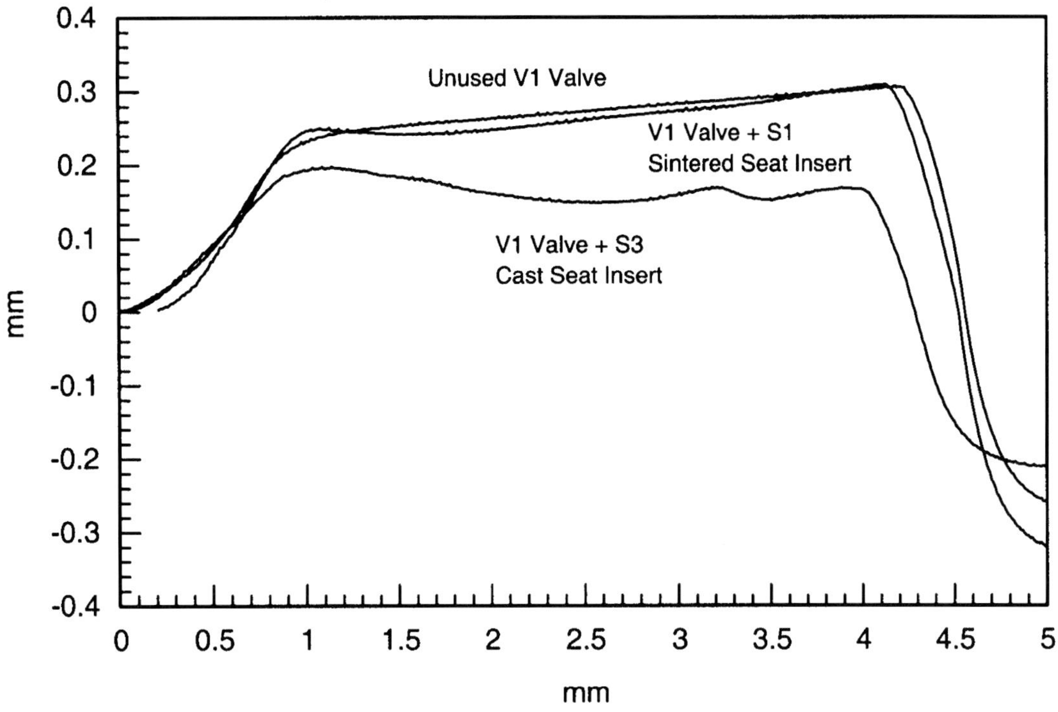

Fig. 4.2 Inlet valve profile traces

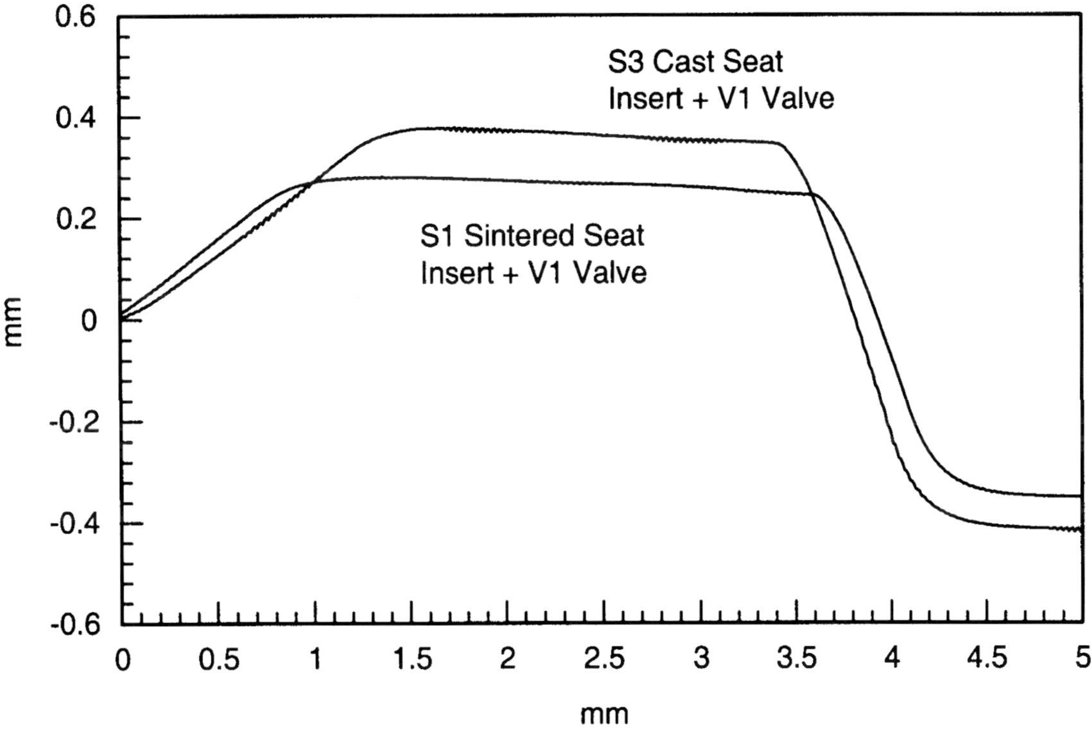

Fig. 4.3 Inlet seat insert profile traces

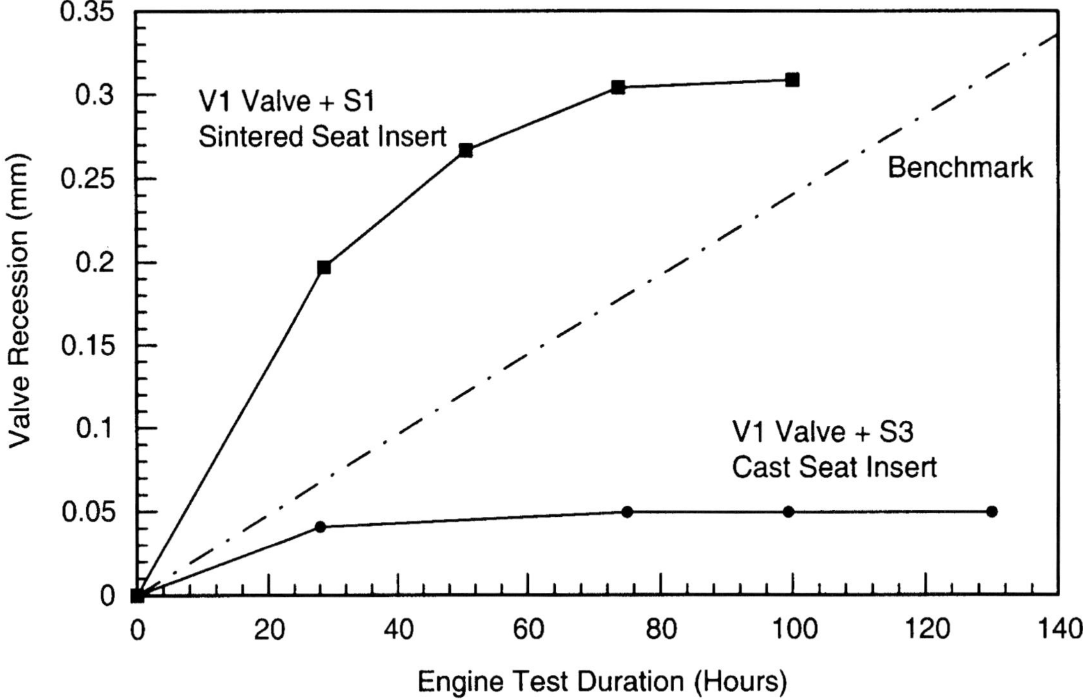

Fig. 4.4 Valve clearance data from engine dyno tests

Profilometry clearly provides data that compare well with valve clearance data taken during engine tests. It also provides an indication of relative valve and seat insert wear rather than just giving an overall figure.

4.2.3 Visual rating

On the inlet valve shown in Fig. 4.5 it is just possible to see the wear scar on the seating face. It can also be seen that the deposits on the valve head compare well with those characterized by Esaki *et al.* [1] (see Fig. 3.11).

In some cases, when using optical microscopy to examine the valves, a series of circumferential ridges and valleys were observed around the axis of the seating faces (see Fig. 4.6). These correspond to the 'single wave formation' described by Van Dissel *et al.* [2]. The seating face of an unused valve is shown in Fig. 4.7 for comparison.

When analysing some of the seat inserts, evidence was found of scratches in the radial direction (see Fig. 4.8) similar to those described by Van Dissel *et al.* [2]. Circumferential grooves and pitting were also observed. Profiles taken indicated that in some tests uneven wear of seat inserts was occurring.

The results of the evaluation of inlet valves and seat inserts from engine dyno tests indicate that they provide a means of comparison and validation for later bench test work. Wear features observed on both valve and seat insert seating faces correspond well with those described in the literature.

Fig. 4.5 Inlet valve (valve material V1 run against insert material S3)

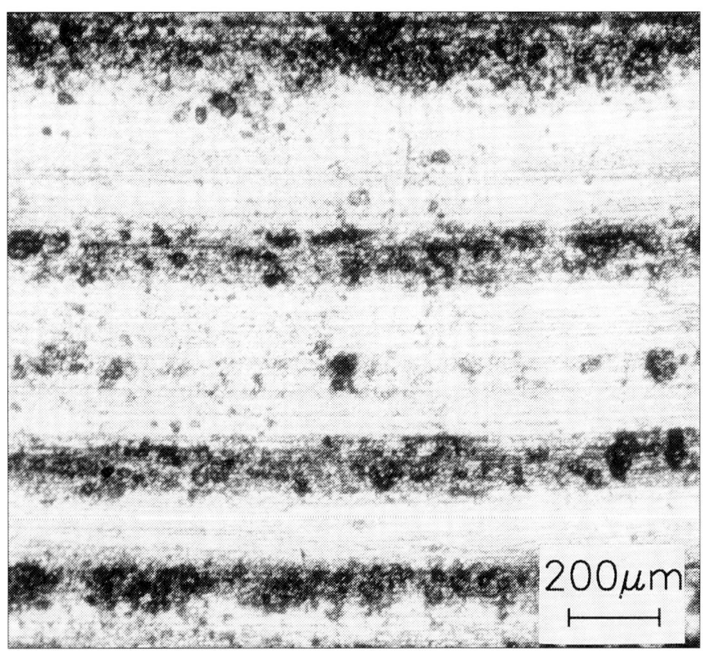

Fig. 4.6 Inlet valve seating face (valve material V1 run against seat insert material S3 in a high-speed engine test); major seat diameter is towards bottom of figure

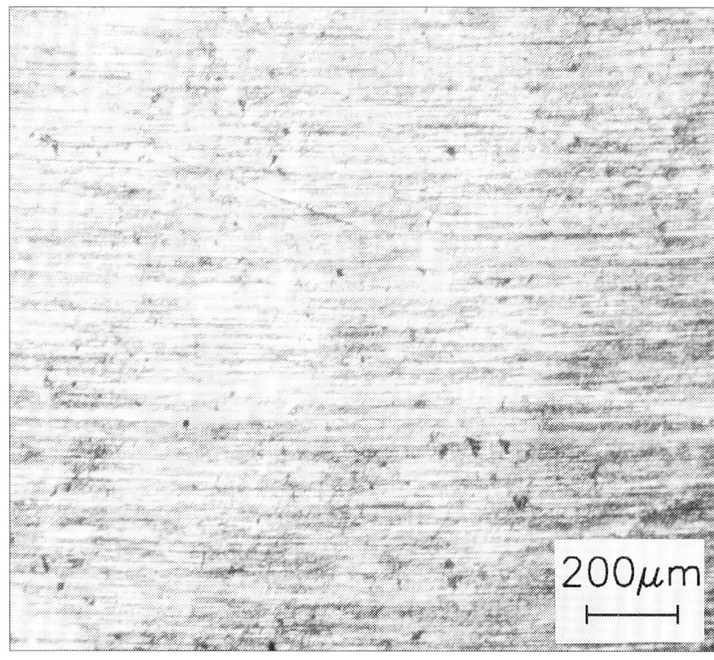

Fig. 4.7 Unused inlet valve seating face (valve material V1); major seat diameter is towards bottom of figure

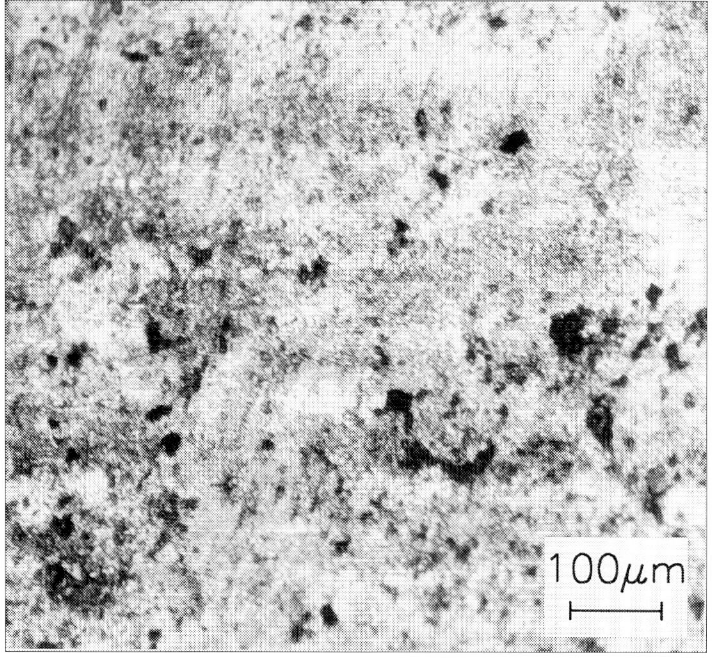

Fig. 4.8 Inlet seat insert seating face (valve material V1 run against seat insert material S1 in a high speed engine test); major seat diameter is towards top of figure

4.3 Lacquer formation on inlet valves

After running a full load durability test on a turbocharged diesel engine, it was found that the seating faces of the inlet valves were coated with what appeared to be a dark lacquer. Inlet valves from a similar test run at part load, in which exhaust gas recirculation (EGR) was able to operate, revealed no presence of lacquer.

Lacquer build-up presents a serious problem because a piece breaking away from the seating face can create a channel through which hot gases are able to escape. This causes guttering of the seating face which eventually leads to valve failure. Failures of this type had been observed in similar full load durability tests on this engine.

The objective of this work was to investigate diesel engine inlet valve deposits and lacquer formation and propose some reasons for the appearance of lacquer on the inlet valves.

4.3.1 Valve evaluation

Valves from both the full load test and the part load test with EGR were evaluated. Valves from each test are shown in Fig. 4.9.

The deposit formation on the valve that underwent the part load test with EGR was black and oily around the valve stem fillet area turning into ash approaching the valve seating face. The valve seating face itself was clear of any deposit. The deposit on the valve that underwent the full load test, however, was virtually all ash and the lacquer formation had visibly dulled the appearance of the valve seating face.

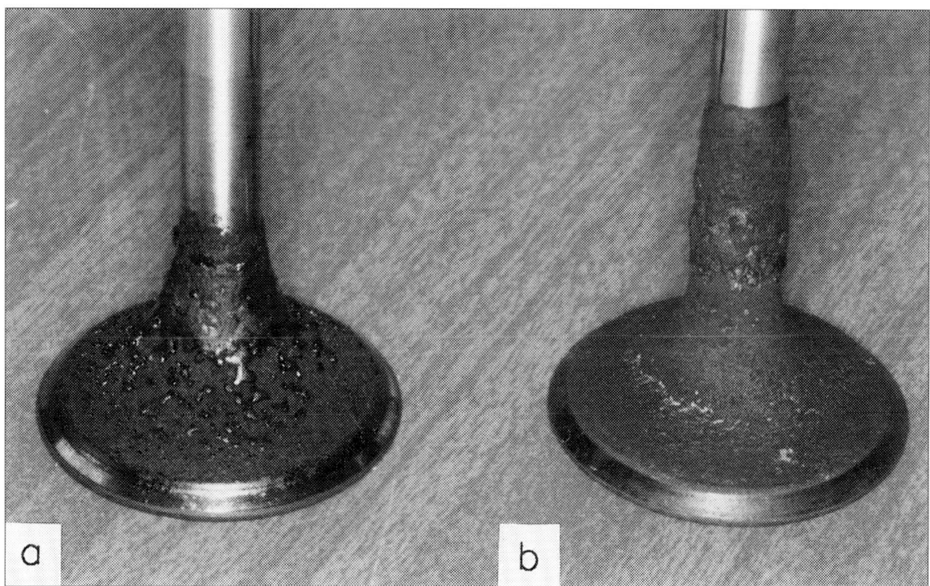

Fig. 4.9 Accumulated deposits on inlet valves: (a) part load test with EGR – no lacquer present on valve seating face; (b) full load test – lacquer present on valve seating face

4.3.2 Discussion

The deposit formation on the valve from the part load test compares well with that characterized by Esaki *et al.* [1], see Fig. 4.10.

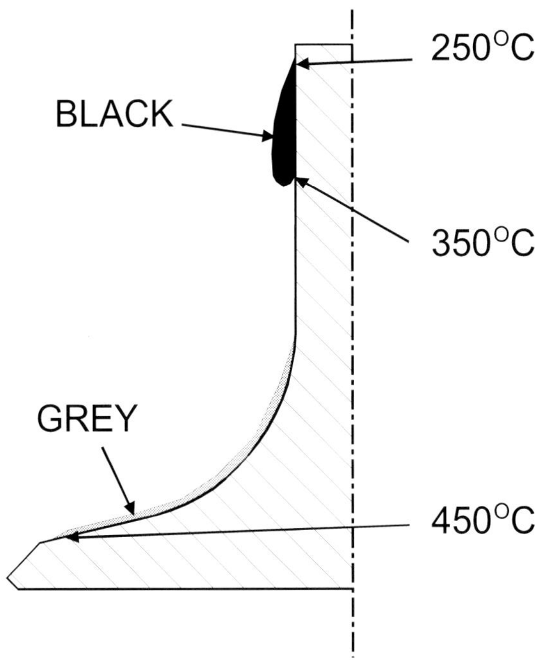

Fig. 4.10 Cross-section of accumulated deposits on diesel engine inlet valves as characterized by Esaki *et al.* [1]

The high ash content of the deposit found on the valves from the full load test clearly indicates that the temperatures were higher than the 350 °C threshold for such deposit formation described above. The temperature of the valve in the part load test was clearly considerably lower, however, hence the more characteristic oily, black deposit formation.

The deposits found on the valves from both tests are clearly composed of lubricating oil that leaked through the valve stem seals onto the valve stem fillets. Lacquer formation is said to occur as a result of oxidation of lubricating oil within the engine [3]. Schilling [4] observed that the unfavourable operating condition in engine parts where lacquer formation was likely to occur was high temperature. This indicates that the lacquer formation on the valves in the full load test could have been caused by the high temperatures experienced at the valve seating face, leading to oxidation of lubricating oil present as a result of valve stem leakage. A reduction in heat transfer from the valve head would have occurred as a result of the lacquer formation, further increasing the valve head temperature and increasing the probability of the lacquer build-up progressing.

The fact that no lacquer was found on the valves from the part load tests can be explained by the lower temperatures experienced. Also, during the part load test when the EGR was in operation, hard combustion particles would have been recirculated in the exhaust gases. These would have provided an abrasive wear medium between the valve seating face and the valve seat insert, further reducing the likelihood of any deposit formation.

4.4 Failure of seat inserts in a 1.8 litre, DI, diesel engine

The engine under consideration in this investigation was a 1.8 litre, DI, naturally aspirated diesel engine with a direct acting cam. This engine is an upgraded version of the 1.8 litre, IDI, diesel engine considered previously. One of the major design changes incorporated was the use of direct fuel injection rather than indirect injection, which required the inclusion of holes in the cylinder head between the inlet and exhaust seat ports to accept the fuel injector (see Fig. 4.11).

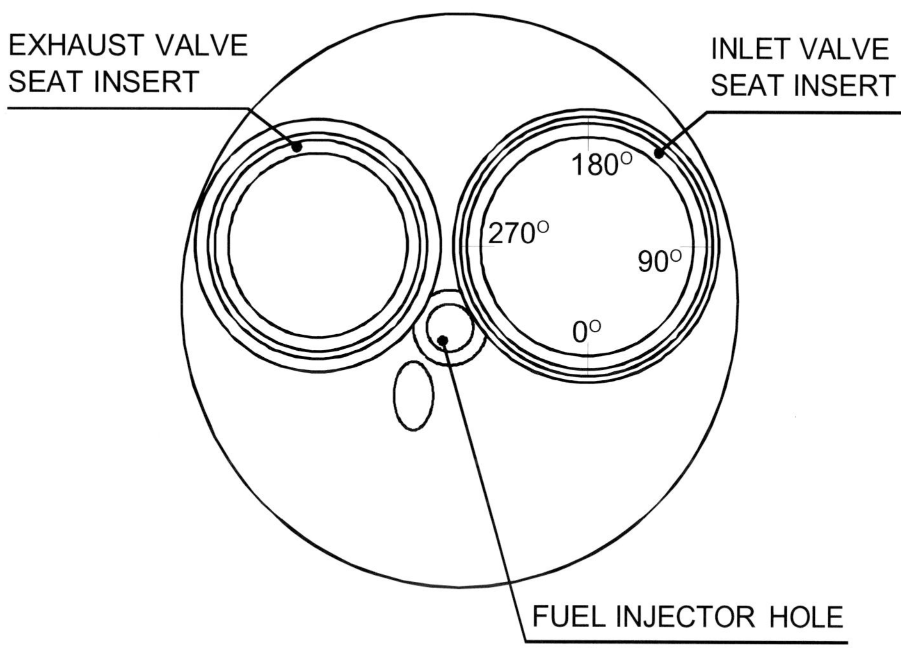

Fig. 4.11 Position of profiles taken on 1.8 l DI inlet seat insert

It was found that on start-up after durability testing, the inlet valves were not seating correctly and consequently sealing was not achieved at the valve/seat insert interface. As a result, pressure was being lost from the cylinder. After the engine had warmed up the valves began to seat correctly. However, on starting the engine again the valves did not seat correctly.

It was hypothesized that this was a wear problem. The seat inserts were deforming (elastically) when hot and being worn unevenly. On cooling they were returning to their original shape. On restarting the engine the inlet valves were not seating correctly as the insert seating faces had been left severely out of round due to the uneven wear. The objective of this work was to establish whether it was a wear problem combined with elastic deformation of the seat inserts or whether there was another reason for the valve/seat insert failures.

4.4.1 Inlet seat insert wear

In order to establish whether, or how much, wear had occurred on the seating faces of the inlet seat inserts, profiles were taken of an inlet seat insert at four positions, each 90 degrees apart, as shown in Fig. 4.11. The seat insert was removed from the cylinder head prior to the profiles being taken. Examples of the resulting plots are shown in Fig. 4.12. The seating face widths were measured to see how much wear had occurred relatively at the four positions (for widths see Fig. 4.13).

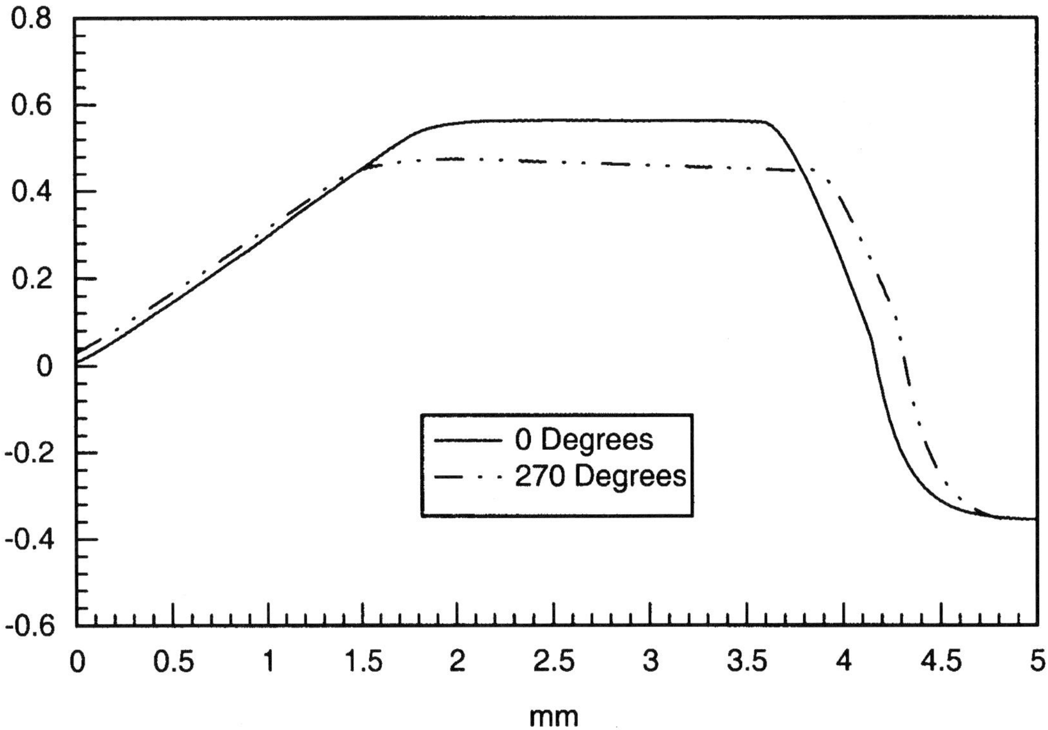

Fig. 4.12 Profile traces taken on 1.8 l DI inlet seat insert

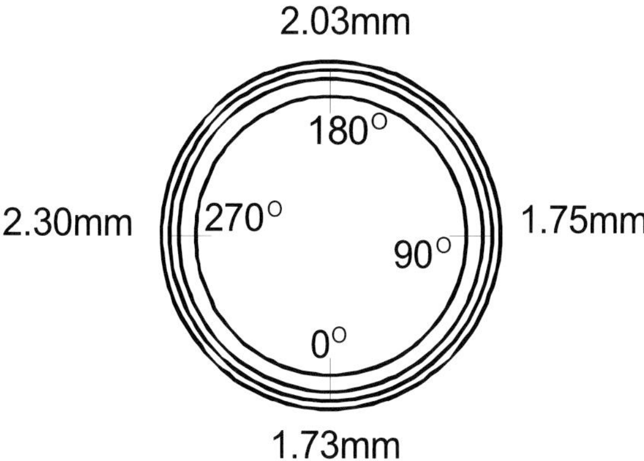

Fig. 4.13 Seating face widths at four positions around inlet seat insert

The measurements clearly indicated that there was uneven wear around the seating face which fitted in with the hypothesis outlined above. The maximum wear was found to have occurred in the proximity of the hole for the fuel injector in the centre of the cylinder head.

Photographs were taken of the seating face of the seat insert (using an optical microscope) that revealed some evidence of indentations in the radial direction, see Fig. 4.14. This indicated that wear resulted from valve head flexure rather than valve rotation.

Fig. 4.14 A 1.8 l DI inlet seat insert seating face (valve material V1 run against seat insert material S3 in a high-speed engine test); major seat diameter is towards top of figure

4.4.2 Deposits

On examining the cylinder head it was found that a thick, black, oily deposit had built up on the inside of the inlet valve seat inserts. As shown in Fig. 4.15, the deposit was also found on the cylinder head material just below the seat insert and on the base of a seat insert removed from the cylinder head. It was noted that the deposit build-up had occurred in the proximity of the hole positioned in the centre of the cylinder head for the fuel injector.

The deposit found on the base of the inlet seat insert could have indicated either that it was not inserted correctly or that it moved during the engine test. Movement could have resulted from non-uniform deformation of the cylinder head and seat insert. This could have been as a result of the fuel injector hole in the centre of the cylinder head.

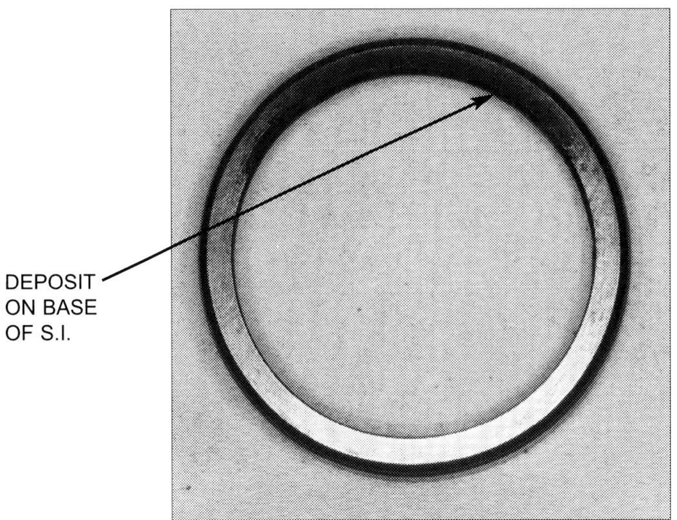

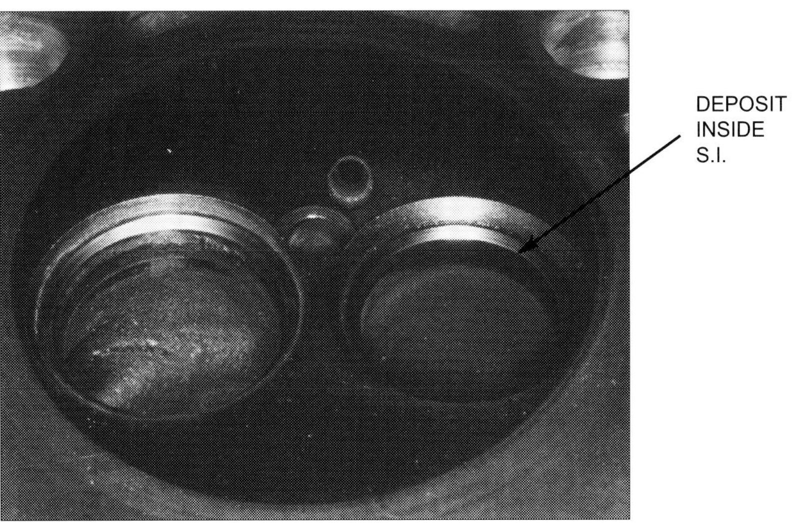

Fig. 4.15 Deposit build-up on 1.8 l DI inlet seat insert

4.4.3 Misalignment of seat insert relative to valve guide

In removing the inlet seat insert from the cylinder head, the cylinder head material around the seat insert was milled away to gain access to the seat insert. The milling machine was centred on the valve guide. During the milling process, it was noted that when the cutter was through to the seat insert on one side, cylinder head material was still present on the other, as shown in Fig. 4.16.

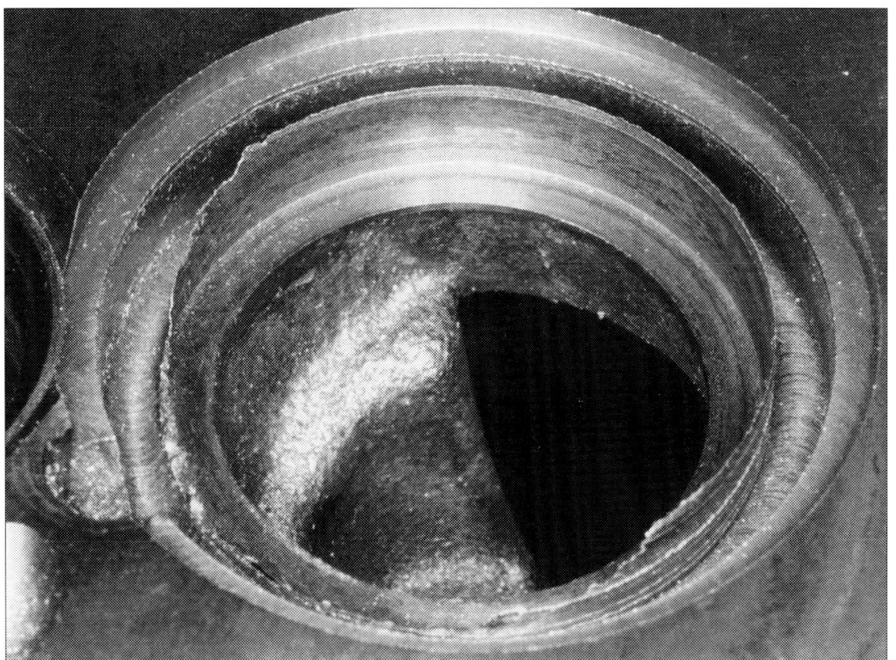

Fig. 4.16 Evidence of seat insert misalignment relative to the valve guide (1.8 l DI)

It was apparent that on insertion, the seat insert had been misaligned relative to the valve guide. This would have been remedied when the grinding of the seating faces on the seat insert occurred, as the grinders are lined up with the valve guide. However, one side of the seat insert was thinner than the other as a result of the initial misalignment. The thinner section was in the proximity of the hole for the fuel injector between the inlet and exhaust ports. This could have affected the heat transfer from the valve head into the cylinder head in this region, which could have led to deformation of the seat insert.

4.4.4 Inlet valve wear

It was thought that if the problem was related to wear, the inlet valve seating faces may exhibit wear of a similar magnitude to that of the inlet seat inserts.

In order to establish whether or how much wear had occurred on the seating face of the valves, profiles were taken of a valve seating face at four positions, each 90 degrees apart. The resulting plots are shown in Fig. 4.17. The similarity of the profiles obtained at the four different positions indicated that if wear had occurred it was evenly distributed

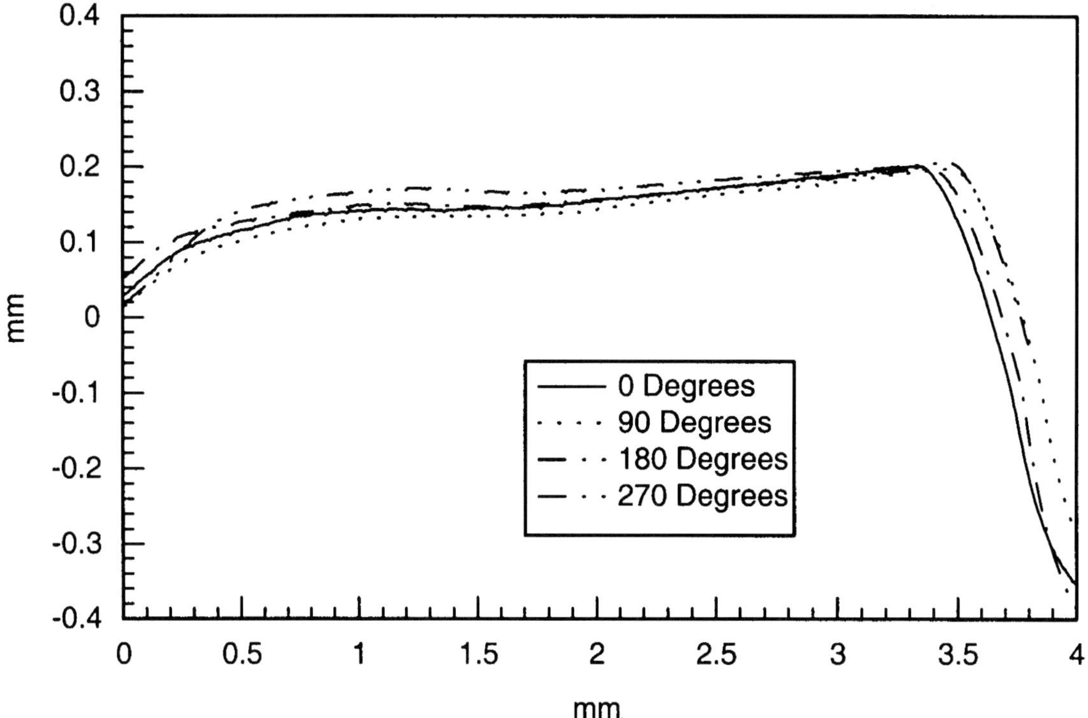

Fig. 4.17 Profile traces taken on 1.8 l DI inlet valve

around the circumference. There was a slight dip in the profiles, which may indicate that some wear occurred. It is possible that this was the point on the valve at which it made contact with the seat insert. Although even valve wear may be expected as a result of rotation, it was clear, as hypothesized, that the problem was related to seat insert wear.

4.5 Conclusions

1. Three failures of valve/seats from dyno tested automotive diesel engines have been analysed.
2. Techniques such as optical microscopy and profilometry have been established for analysis.
3. The following parameters have been identified as influencing valve recession:
   - valve and seat insert material properties;
   - valve and seat insert misalignment (caused by cylinder head deformation as a result of non-uniform cooling);
   - deposit formation on the seating face of a valve.
4. Wear data and surface appearance of failed components were available for comparison and validation with later bench testing.

The work has also further emphasized the unique nature of each valve recession problem. Valve recession is certainly not purely a materials selection issue; it is also affected greatly by the valve train and cylinder head design and manufacturing tolerances.

4.6 References

1. **Esaki, Y., Ishiguro, T., Susuki, N.,** and **Nakada, M.** (1990) Mechanism of intake valve deposit formation: Part 1 – Characterization of deposits, SAE Paper 900151, *SAE Trans.*, **99**.

2. **Van Dissel, R., Barber, G.C., Larson, J.M.,** and **Narasimhan, S.L.** (1989) Engine valve seat and insert wear, SAE Paper 892146.

3. **Denison, G.H.** and **Kavanagh, F.W.** (1955) Recent trends in automotive lubricating oil research, Section 6/C, Preprint 1, Proc. Fourth *World Petroleum Congress*, Rome.

4. **Schilling, A.** (1968) *Motor oils and engine lubrication*, Scientific Publications (GB) Ltd.

Chapter 5

Valve and Seat Wear Testing Apparatus

5.1 Introduction

Dynamometer engine testing is often employed to investigate valve wear problems. This is expensive and time consuming and does not necessarily help in finding the actual cause of wear. Since valve wear involves so many variables, it is impossible to confirm precisely individual quantitative evaluations of all of them during such testing. In addition, the understanding of wear mechanisms is complicated by inconsistent patterns of valve failure. For example, failure may occur in only a single valve operating in a multi-valve cylinder. Furthermore, the apparent mode of failure may vary from one valve to another in the same cylinder or between cylinders in the same engine. Each case, therefore, has to be painstakingly investigated, the cause or causes of the problem isolated, and remedial action taken. In order to isolate the critical operating conditions and analyse the wear mechanisms in detail, simulation of the valve wear process must be used. This has the added benefits of being cost effective and saving time.

This chapter details the requirements of valve and seat wear test apparatus, wear test methods, and extant valve wear test rigs. It then goes on to describe work undertaken at the University of Sheffield to design and build experimental apparatus that would be able to simulate the loading environment and contact conditions to which the valve and seat insert are subjected in an engine. The apparatus was to be used in future bench test work intended to isolate parameters critical to the valve recession problem.

5.2 Requirements

The review of literature and the failure diagnosis carried out indicated that there were a number of critical parameters which affect the rate of valve and seat insert wear: combustion load influences the frictional sliding; high temperatures affect the wear mechanism and the formation of deposits; misalignment leads to uneven seating loads; deposit formation affects heat transfer; and rotation plays a role as yet not understood. The major requirements of valve and seat wear test apparatus could, therefore, be listed as:

- combustion loading;
- impact load on closing;
- valve misalignment;

- valve rotation;
- temperature control.

5.3 Wear test methods

A number of different wear tests developed to evaluate material wear are compared in Table 5.1. They are listed in order of increasing complexity.

Table 5.1 Wear test methods [1]

Test method	*Type of test*	*Test conditions*	*Measured quantity*
Crossed cylinder (ASTM G 83-83)	Adhesive wear	High-contact stress High-sliding velocity No lubrication	Weight loss
Block on ring (ASTM G 77-83)	Adhesive (sliding)	High-contact stress Sliding speed High temperature No lubrication	Weight loss Friction
Thrust washer	Adhesive/abrasive	High-contact stress Sliding speed High temperature No lubrication	Weight loss Wear depth Wear profile Cycles to failure
Bench test rigs	General	Valve gear lube Speed Temperature Spring load Seating velocity	Oil residue analysis Wear depth Wear profile
Motorized or fired engine tests	General	Engine operating conditions Speed Torque	

Basic wear tests are standard tests. They allow close control of the test conditions such as loads, environment, and dimensions. The 'crossed cylinder' and 'block on ring' tests are generally used to evaluate adhesive wear resistance. Thrust washer tests, at high temperatures, can be used to simulate valve seat insert wear conditions. Bench test rigs simulate actual valve gear operating conditions. However, they only allow limited control over operating conditions. Engine tests can be motorized fixtures, firing engine dynamometer tests, or fleet tests with actual vehicles. These are expensive and time consuming, however, and it is difficult to isolate the actual cause of any wear that occurs.

5.4 Extant valve and seat wear test rigs

A review of extant valve wear test rigs was carried out in order to look at how wear investigations or material ranking could be achieved using the various testing methods outlined in Section 5.3.

The test rigs reviewed could be split into three different types:

1. static test rigs used to measure deflection;
2. wear test rigs using material specimens;
3. wear test rigs utilizing actual valves and seat inserts.

Brief notes on each of the rigs reviewed can be found in Table 5.2 and diagrams of some of the rigs are shown in Fig. 5.1. Table 5.3 gives a summary of the engine operating parameters that can be simulated using the rigs reviewed.

Table 5.2 Details of the valve/seat wear test rigs reviewed

Reference	Notes
Zinner [2]	Rig was used to study the effect of seat insert distortion on the wear of diesel engine inlet valves. Static rig in which actual seating condition was simulated. Seat insert was forced from side to artificially deform. Valve stem protrusion was measured at different air pressures applied to the valve head. It was found that seat insert distortion increased valve stem protrusion. It was concluded that partial contact of the mating faces increased valve seat slide, causing excessive valve seat wear.
Pope [3]	Rig was constructed to check the validity of Pope's theoretically derived wear factor for inlet valve seats. Valve stem protrusion was measured at different oil pressures applied to the valve head. Results showed that the magnitude of the theoretical deflection was of the correct order.
Matsushima [4]	Rig was used to study the wear rate of valve face and seat inserts at elevated seat temperatures. Several powder metal seat insert materials were tested. The rig was able to move the valve up and down to simulate the opening and closing motion. The valve was also rotated. The valve was heated using a gas burner. Not clear how or whether the valve was loaded during the opening and closing cycle.
Narasimhan and Larson [1]	The test rig was used for wear testing valve seat and seat insert materials at high temperatures. The valve head was heated by a gas burner and the temperature of the valve head, seat and seat insert were monitored and controlled during the test. Valve seating velocity and the seating loads were controlled using a servo-hydraulic actuator. Tests were performed to a fixed number of cycles or until failure occurred.
Blau [5]	The test rig was used to simulate repetitive impact and seating of a valve on its seat. Rectangular ceramic coupons were used and a spherically tipped hammer produced the impact. The rig was able to heat and lubricate the test specimens.
Hofmann *et al.* [6]	Hot wear tester designed to assess the hot wear properties of powder metal alloys used in seat inserts. The tester was able to control the test variables of load, temperature, atmosphere and speed. The contact geometry was a rotating cylinder on a stationary block. A positive correlation was found between wear loss data and engine test results.
Nakagawa *et al.* [7]	The wear test rig was used to test hard surfacing alloys for internal combustion engine inlet valves. Valve face wear depth was measured for different valve face temperatures.
Fujiki and Makoto [8]	Valve wear test rig utilized actual valve gear. The valve was heated using a gas burner. It is not clear how or whether the valve was loaded. A degree of success was indicated in correlating valve seat wear measurements from bench testing with engine dynamometer testing.
Malatesta *et al.* [9]	Test rig utilized a hydraulic actuator to compressively load a valve against its seat insert. A gas burner was used to heat the valve. It was possible to produce various conditions of misalignment (angular and lateral). The valve was prevented from rotating in the rig. A load cell was used to monitor the magnitude of the seating load, and in conjunction with a control loop the desired load was maintained. Sample and system temperatures were monitored using thermocouples. The test rig was able to achieve seating velocities of approximately 250 mm/s, load the valve to a maximum of 37 810 N and produce valve seat temperatures up to 816 °C. Studies were made of the relationship between the number of cycles and both wear depth and area. As a means to validate the results a comparison was made with several heavy duty diesel valves.

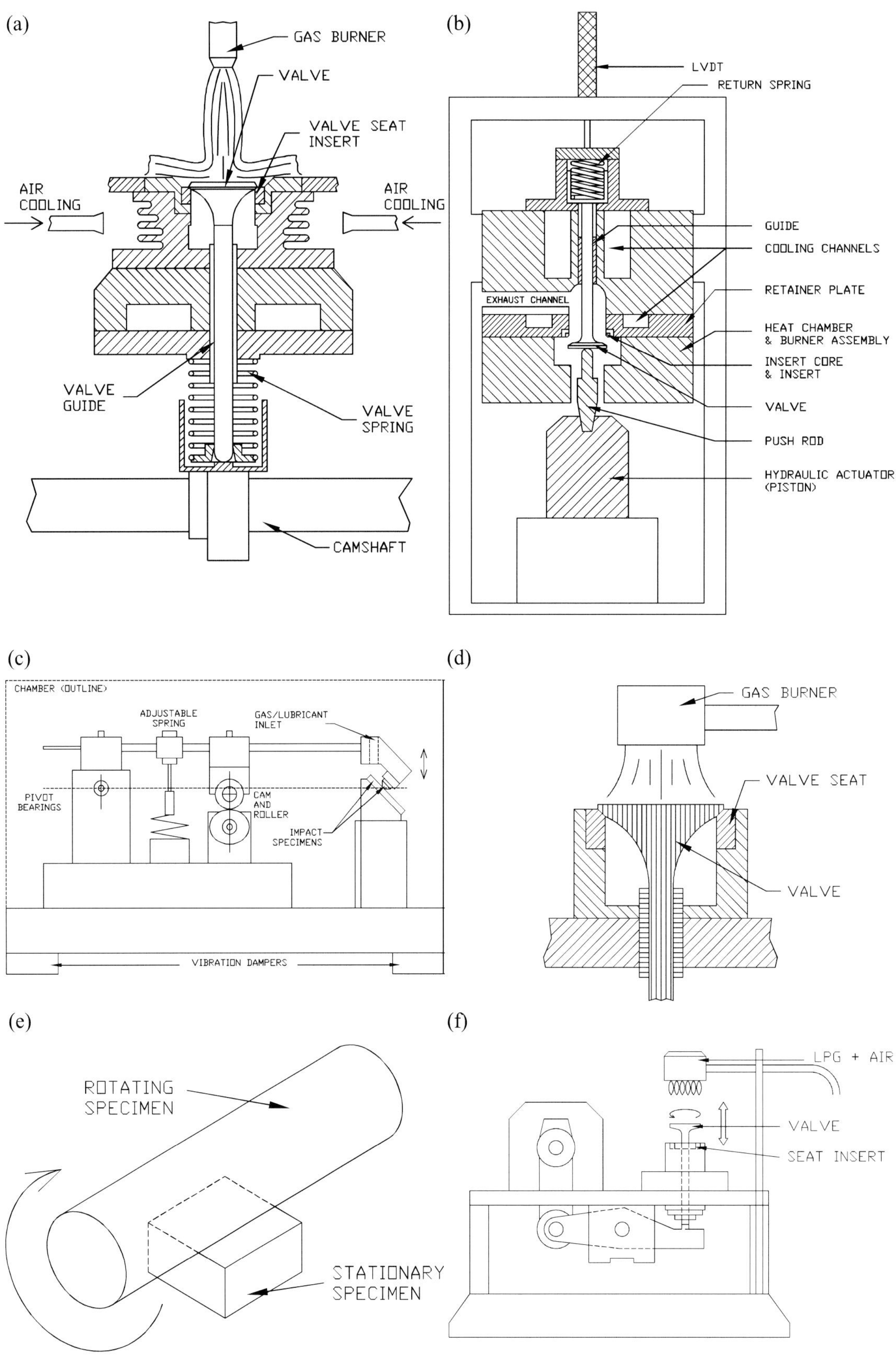

Fig. 5.1 Examples of the test rigs reviewed: (a) Fujiki and Makoto [8]; (b) Malatesta *et al.* [9]; (c) Blau [5]; (d) Nakagawa *et al.* [7]; (e) Hofmann *et al.* [6]; (f) Matsushima [4]

Table 5.3 Summary of engine operating parameters that can be simulated by the test rigs reviewed

Reference	Rig type	Impact load on valve closing	Combustion loading	Valve rotation	Valve misalignment	Temperature control
Zinner [2]	Static		•		•	
Pope [3]	Static		•			
Narasimhan and Larson [1]	Wear	• (components)		•		•
Hofmann *et al.* [6]	Wear (specimens)	•	•			•
Matsushima [4]	Wear (components)					•
Nakagawa [7] *et al.*	Wear (components)					•
Fujiki and Makoto[8]	Wear (components)	•		•		•
Blau [5]	Wear (specimens)	•		•	•	•
Malatesta [9] *et al.*	Wear (components)	•	•		•	•

It was found that wear test rigs using material specimens provided good control over test conditions. However, without using actual components it was difficult to recreate exact contact conditions and it was, therefore, impossible to simulate wear features found on components taken from the field. Such test rigs could not be used for the study of wear mechanisms, but were suitable for ranking of valve and seat material performance.

Wear test rigs utilizing actual valve train components were found to provide the best means of replicating wear features and studying wear mechanisms. While they offered less control over operating conditions than rigs using material specimens, it was still possible to isolate important test parameters.

5.5 University of Sheffield valve seat test apparatus

Having established the requirements of the apparatus and the methods available for valve gear wear testing (see Section 5.3), and using the results of the review of extant test rigs (see Section 5.4), it was decided that, in order to realistically simulate the contact conditions to which a valve and seat are subjected in an engine, the best approach would be to design apparatus that utilized actual valve and seat inserts.

In order to accurately replicate impact load on valve closure it was clear that actual valve gear would have to be used. However, it was determined that producing the combustion loading conditions would be difficult using camshaft and valve train components and that in order to investigate the effect of both it would be necessary to design two test rigs: one designed to fit in a hydraulic test machine, to study the effect of combustion loading while approximating valve dynamics; and a motorized cylinder head utilizing actual valve gear, to study the effect of impact on valve closure without the application of combustion loading.

Different configurations were considered for the two rigs. The final design, however, for the combustion loading rig was mainly determined by the test facilities available for applying the required loading. It was decided to base the impact rig on a motorized cylinder head.

5.5.1 Hydraulic loading apparatus

5.5.1.1 Design

The test rig (shown in Fig. 5.2) was designed to be mounted on a hydraulic fatigue testing machine (as shown in Fig. 5.3). The hydraulic actuator on the machine is used to provide the combustion loading cycles required and acts to 'close' the valve. A spring returns the valve to the 'open' position. The control system allows loading or displacement waveforms and their amplitude to be set, as well as the frequency of the loading or displacement. A built-in load cell and linear variable displacement

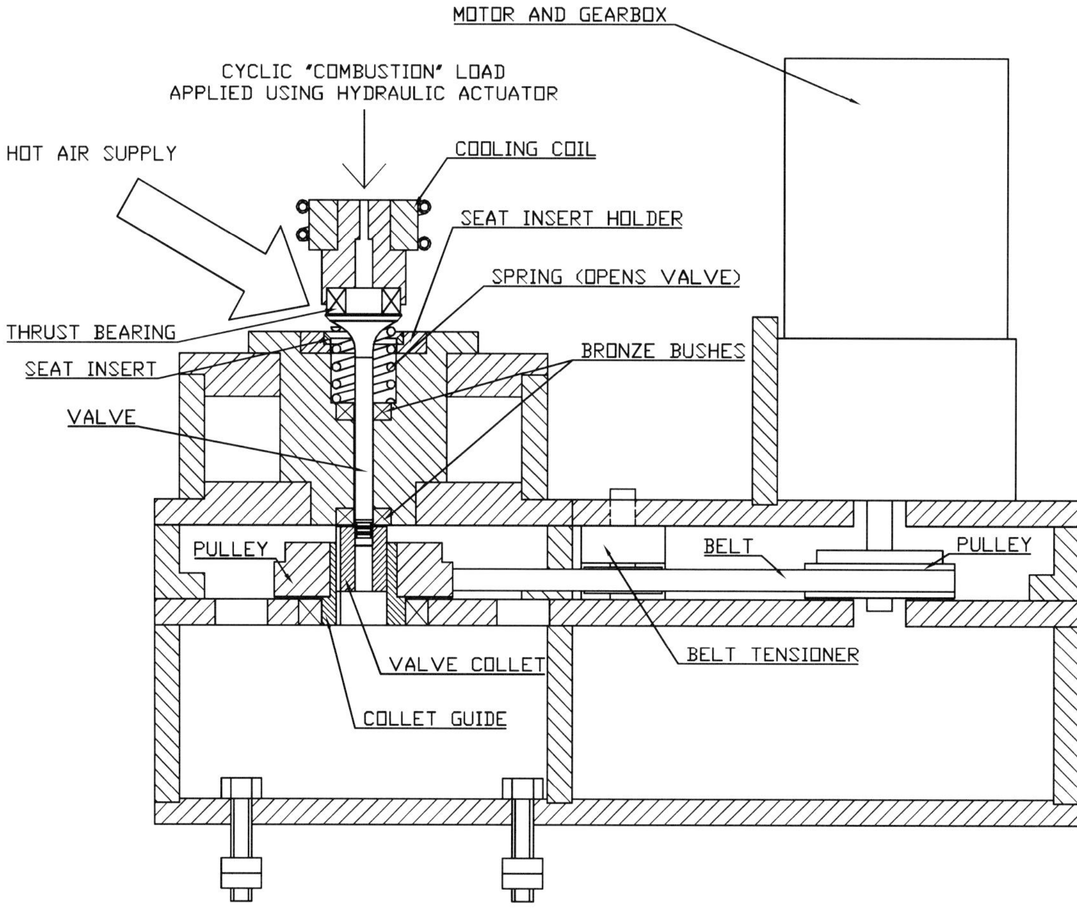

Fig. 5.2 Hydraulic loading apparatus (Reprinted with permission from SAE paper 1999-01-1216 © 1999 Society of Automotive Engineers, Inc.)

Fig. 5.3 Test rig mounted in a hydraulic test machine

transducer (LVDT) provide load and displacement measurements. A counter records the number of loading cycles.

A removable seat insert holder is used to mount the seat insert in alignment with the valve. An additional holder was designed in which the seat insert was slightly off-centre in order to misalign the valve relative to the seat insert. A valve guide of bronze bushes was also built into the rig. The bushes and seat insert holders are designed to accommodate inlet valves and seat inserts from a 1.8 litre, IDI, automotive diesel engine, the geometries of which are shown in Fig. 5.4. The materials not available as seat inserts are made up into specimens as shown in Fig. 5.5. The use of a seat insert holder and an inserted valve guide allows some flexibility in the size of the valve and seat insert being tested.

Heating is provided to both sides of the rig by hot air supplies directed at the valve and seat insert. Temperature measurements are taken using a contact probe placed on the outer edge of the valve head at the top of the seating face. A cooling coil is provided to prevent the load cell from overheating.

Valve rotation is achieved using a motor-driven belt and pulley system as shown in Fig. 5.6. The valve collet, clamped around the valve stem, is able to move up and down in the collet guide while rotating (see Fig. 5.7). The valve is not intended to rotate while seated.

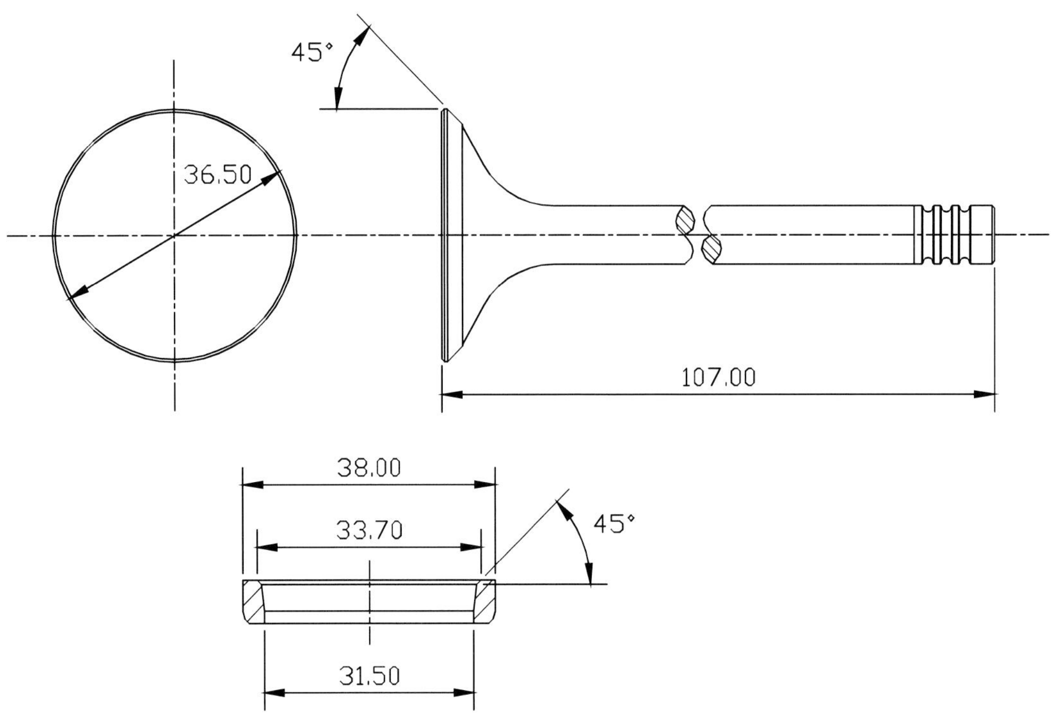

Fig. 5.4 Valve and seat insert geometry (Reprinted with permission from SAE paper 1999-01-1216 © 1999 Society of Automotive Engineers, Inc.)

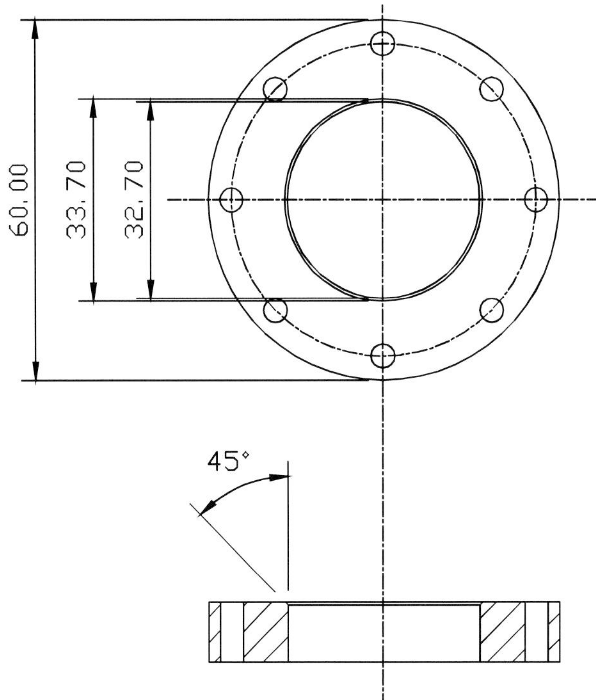

Fig. 5.5 Seat specimen geometry

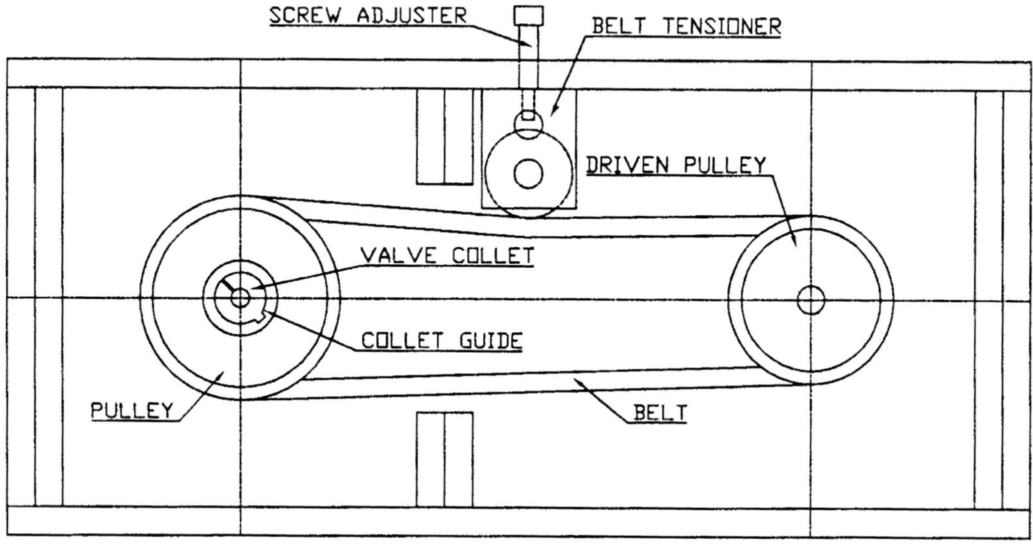

Fig. 5.6 Plan view of valve rotation system

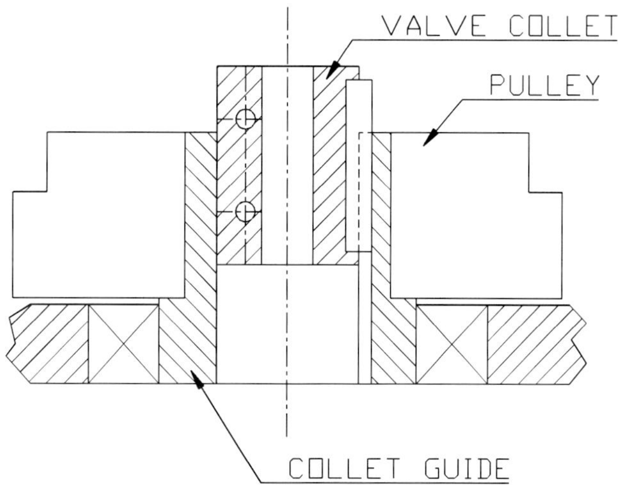

Fig. 5.7 Valve collet and collet guide

Lubrication of the valve/seat insert interface can be achieved via holes in a tube placed around the thrust bearing housing above the valve head (see Fig. 5.8). Flow is controlled using a valve fitted below the lubricant reservoir.

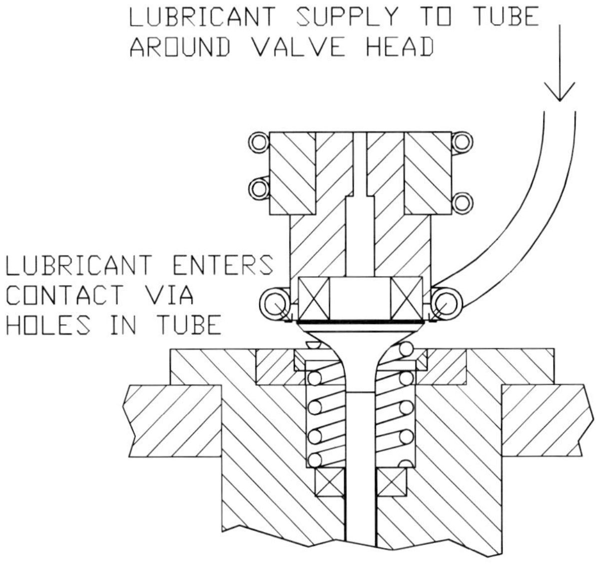

Fig. 5.8 Lubrication of valve/seat insert interface

5.5.1.2 Test methodologies

In order to develop test methodologies with which to investigate the likely causes of valve and seat insert wear and establish experimental parameters, testing was carried out on the hydraulic loading apparatus to evaluate the performance of the hydraulic test machine to which it was mounted.

Two different test methodologies were developed, the first to investigate the effect of the frictional sliding of the valve on the seat insert under the action of the combustion pressure, and the second to investigate the effect on wear of combining this with the impact of the valve on the seat insert during valve closure.

Frictional sliding

The first methodology employed a triangular loading waveform to investigate the effect of the combustion loading on valve and seat insert wear (see Fig. 5.9). The intention was to isolate the frictional sliding between the valve seating face and the seat insert, hence the valve seating face was in constant contact with the seat insert. A triangular waveform was used as this was the closest approximation to the cylinder pressure curve for a compression ignition engine (as shown in Fig. 5.10).

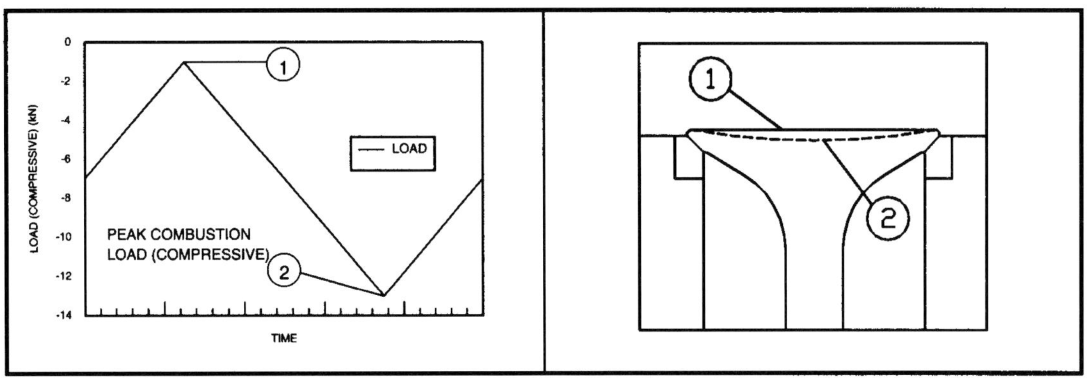

Fig. 5.9 Frictional sliding test methodology

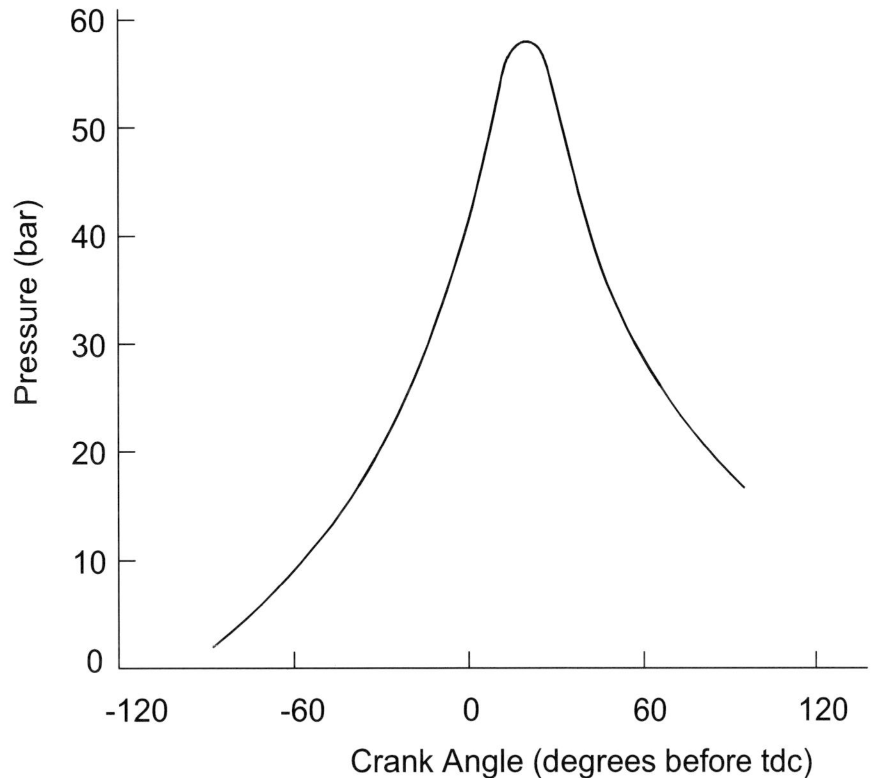

Fig. 5.10 Hypothetical pressure diagram for a compression ignition engine [10]

Impact and sliding

The second test methodology used a sinusoidal displacement waveform (allowing the valve to lift off the seat), to investigate the effect of the impact of the valve on the seat insert as the valve closes, in combination with the combustion loading (see Fig. 5.11). A sinusoidal waveform was used as this was the closest available approximation of the motion of a valve in an engine.

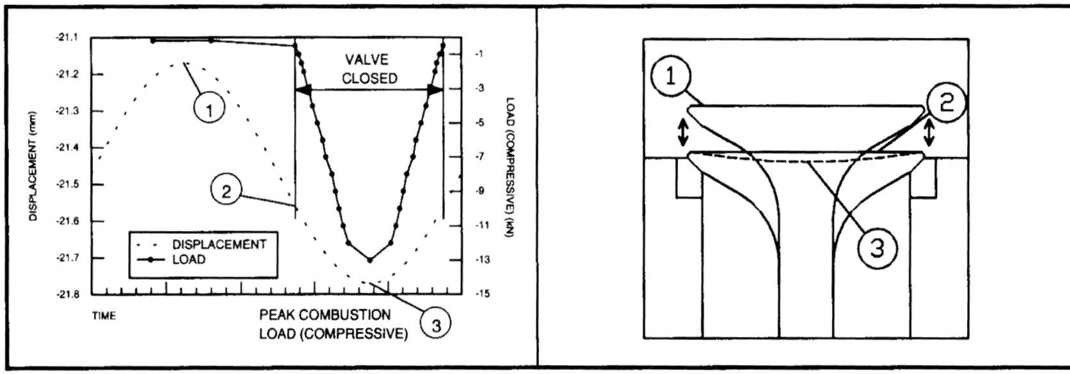

Fig. 5.11 Impact and sliding test methodology

5.5.1.3 Experimental parameters

During initial testing, load, misalignment, rotation, and temperature were varied in order to establish baseline parameters for each of the test methodologies outlined above. Test parameters used are shown in Table 5.4.

Table 5.4 Parameters varied during initial testing

Frictional sliding tests

Combustion load (kN)	Load waveform	Frequency (Hz)	Misalignment (mm)	Valve temp. (°C)	Rotation (r/min)
13–15	Triangular	5–20	0–0.5	R.T.–130	0–1

Impact and sliding tests

Combustion load (kN)	Displacement waveform	Valve lift (mm)	Frequency (Hz)	Misalignment (mm)	Valve temp. (°C)	Rotation (r/min)
13	Sinusoidal	0.6–1	10–15	0–0.5	R.T.–130	0–1

The magnitude of the load to be used, P_p, was calculated by multiplying the maximum combustion pressure, p_p, by the valve head area for an inlet valve in a naturally-aspirated (N/A), 1.8 litre, IDI, diesel engine, see equation (5.1)

$$P_p = p_p \times \pi \times R_v^2 \tag{5.1}$$

where P_p is the peak combustion load (N) and R_v is the radius of valve head (m).

For a N/A, 1.8 litre, IDI, diesel engine, p_p = 13 MN/m^2 and R_v = 18×10^{-3} m.

Therefore, from equation (5.1)

peak combustion load, P_p = 13.2 kN

Other parameters, such as frequency, lift, misalignment, and rotation, were varied in order to optimize the test rig performance at the calculated load. Temperatures were restricted by the maximum temperature to which the load cell within the hydraulic test machine could be exposed. The baseline parameters established for optimum rig performance are shown in Table 5.5.

Table 5.5 Baseline test parameters

Frictional sliding tests						
Combustion load (kN)	Load waveform		Frequency (Hz)	Misalignment (mm)	Valve temp. (°C)	Rotation (r/min)
13	Triangular		20	0.25	R.T.	1
Impact and sliding tests						
Combustion load (kN)	Displacement waveform	Valve lift (mm)	Frequency (Hz)	Misalignment (mm)	Valve temp. (°C)	Rotation (r/min)
13	Sinusoidal	0.6	10	0.25	R.T.	1

5.5.2 Motorized cylinder head

5.5.2.1 Design

The test rig (shown in Fig. 5.12) utilizes a cylinder head from a 1.8 litre, IDI, diesel engine. The cylinder head is bolted to an adjustable bedplate which is mounted on a steel frame. An electric motor (1440 r/min rated speed), housed within the frame, is used to drive the camshaft via a belt and pulley system. This gives the camshaft a rotational speed of approximately 2700 r/min. For each rotation of the camshaft, the inlet valves open and close once. A soft starter is used to provide a gradual increase and decrease of motor torque during start-up and shut-down, respectively. This suppresses explosive start-up conditions and reduces peak loads on drivetrain components. It does not, however, allow the rotational speed of the motor to be varied. The camshaft speed is therefore fixed, but could be varied by using different pulley configurations.

An overhead camshaft configuration is employed in this valvetrain system, with the cams acting directly on flat faced followers. The cams are offset from the valve stem axis (as shown in Fig. 5.13) in order to reduce localized wear on the follower. The use of offset cams, along with split collets, promotes valve rotation.

Removable seat insert holders are used in order to speed up analysis during testing. These are clamped into holes machined in the cylinder head around the inlet ports. A gravity-fed oil drip system is used to lubricate the cam/follower interfaces and the camshaft bearings. A collection tray/shield fitted around the camshaft is used to recycle the lubricant.

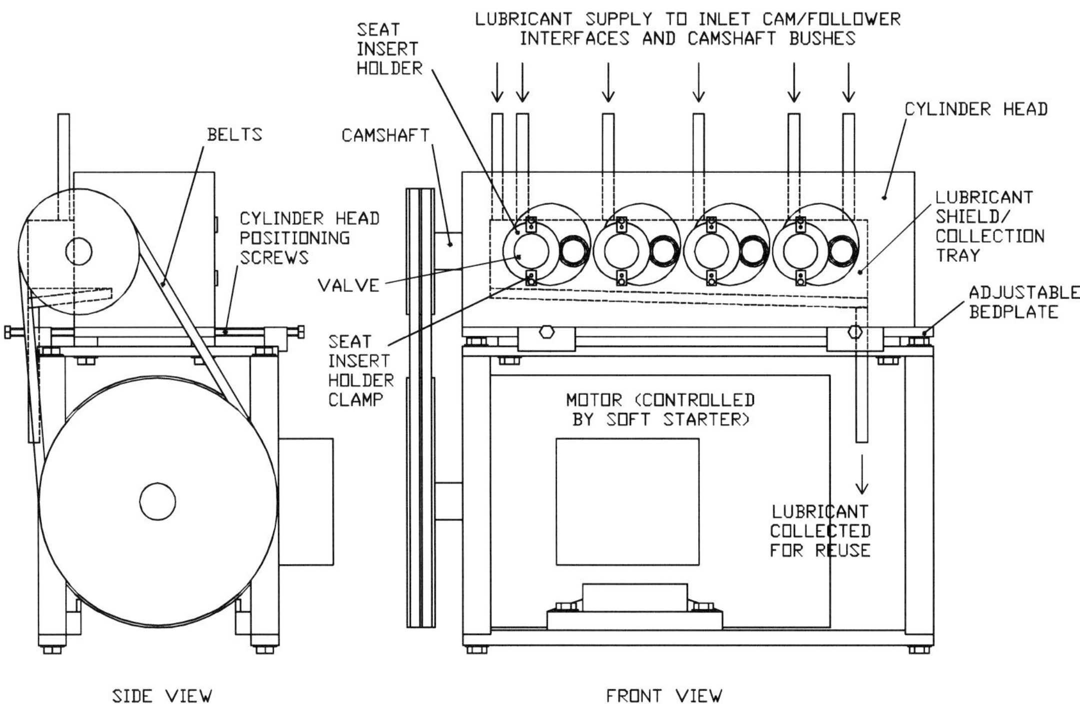

Fig. 5.12 Motorized cylinder head

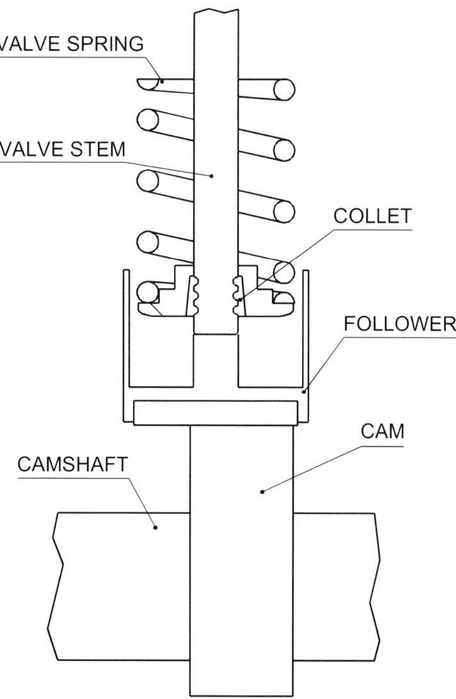

Fig. 5.13 Cam offset from the valve stem axis

With regard to the study of impact on valve closure, the motorized cylinder head has the following advantages over the hydraulic loading apparatus described in Section 5.5.1:

- actual valve dynamics are utilized;
- impact on valve closure is isolated;
- actual valve rotation can be studied;
- valve dynamics can be varied by using different cam profiles.

5.5.2.2 Operation

Development of a test methodology and selection of experimental parameters was more straightforward for the motorized cylinder head. In studying the impact of the valve on the seat at valve closure, only variation of the valve closing velocity was required.

In order to vary the valve closing velocity, the rotational speed of the camshaft could be changed. In order to change the closing velocity of individual valves, the clearance was adjusted by employing seat insert holders of different thickness.

5.5.3 Evaluation of dynamics and loading

In order to assess the extent to which the dynamics of the hydraulic test machine on which the test rig was mounted simulated the dynamics of an inlet valve, it was decided to investigate the dynamics of an inlet valve in a 1.8 litre, IDI, diesel engine with a direct-acting cam and compare them with those for the hydraulic test machine.

5.5.3.1 1.8 litre, IDI, diesel engine

The valve lift curve for the 1.8 litre, IDI, diesel engine was taken directly from the inlet cam lift data (lift versus angle of rotation). The valve velocity, *v*, was then derived by multiplying the gradient of the lift curve by the rotational speed of the camshaft ω, see equation (5.2). An engine speed of 4800 r/min was assumed (engine speed for durability tests) giving a camshaft rotational speed of 2400 r/min.

$$\mathrm{v} = \omega \frac{\mathrm{d}l}{\mathrm{d}\theta} \tag{5.2}$$

where l is the valve lift (mm) and θ is the rotation of camshaft (degrees).

Both valve lift and velocity curves are shown in Fig. 5.14. Ideally the valve should close in the region of constant valve velocity between 145 and 160 degrees of camshaft rotation in order to limit impact stresses.

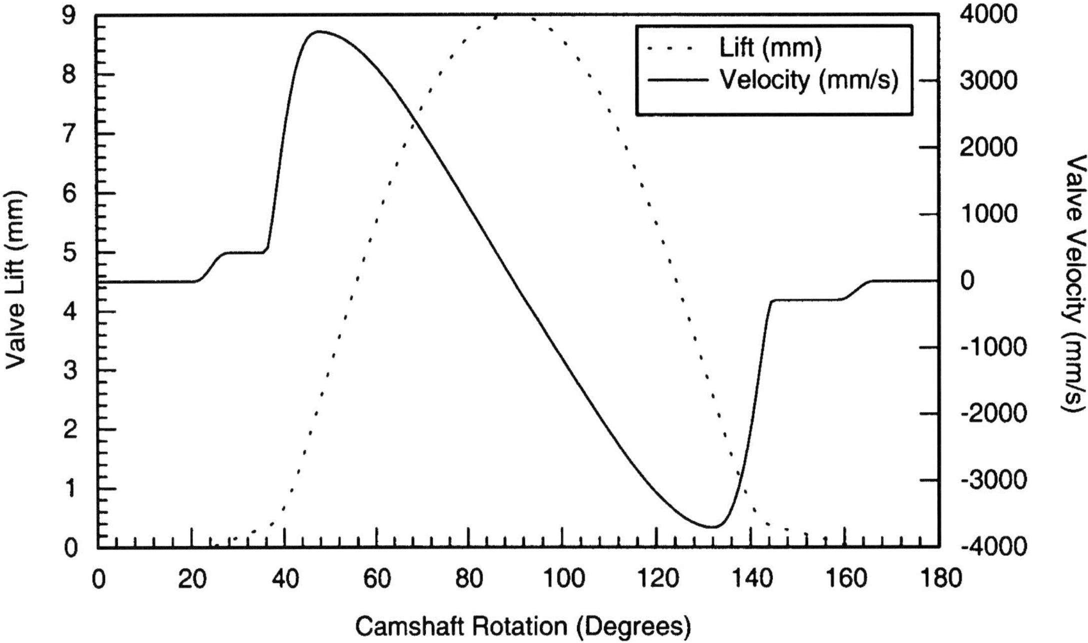

Fig. 5.14 1.8 l, IDI, diesel engine inlet valve lift and velocity

The valve lift and valve velocity curves plotted against cam rotation for the region in which valve closure occurs are shown in Fig. 5.15 (magnification of Fig. 5.14). Valve closure does not occur when the lift is equal to zero as a clearance is introduced between the valve tip and the follower. This allows the engine to tolerate a certain amount of valve recession. The recommended maximum and minimum clearances for this type of engine are shown in Fig. 5.15. The valve closing velocities for the maximum and minimum clearances, also shown in Fig. 5.15, are 375 mm/s and 288 mm/s, respectively. These values are well below the maximum recommended value for valve closing velocity of 500 mm/s given by Stone [10].

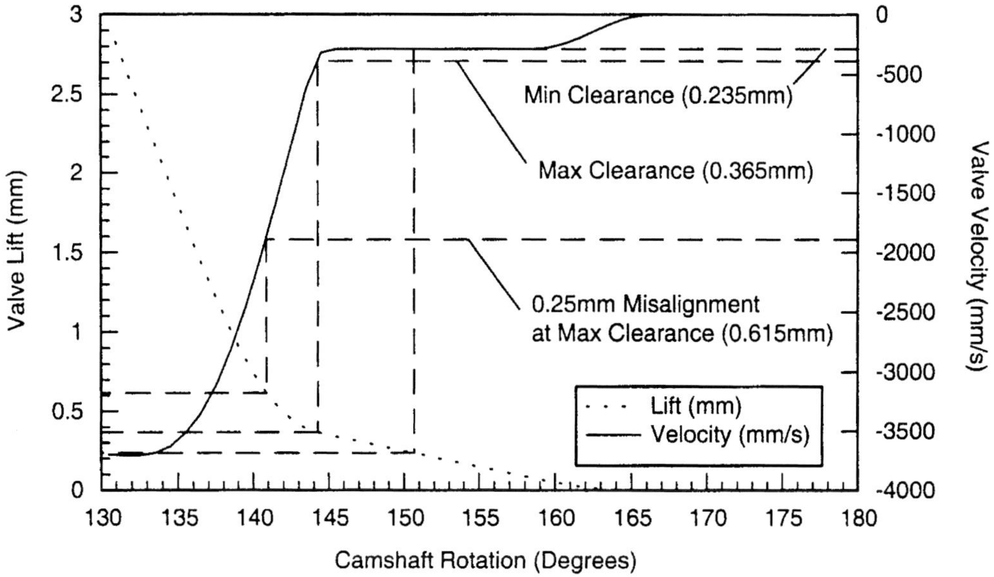

Fig. 5.15 1.8 l, IDI, diesel engine inlet valve lift and velocity at valve closure

Valve misalignment relative to the seat insert will cause the valve to impact the seat insert at a higher valve lift (as shown in Fig. 5.16). For a seating face angle of 45 degrees, the amount of misalignment will equal the increase in valve lift at closure. As shown in Fig. 5.15, a valve at maximum clearance misaligned by 0.25 mm closes off the constant velocity ramp and the closing velocity is increased to 1860 mm/s.

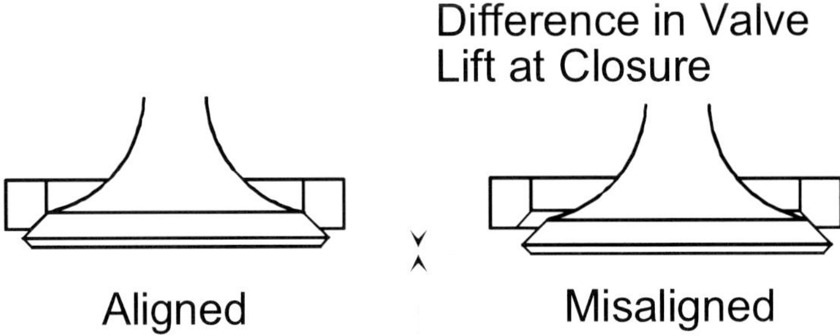

Fig. 5.16 Effect of valve misalignment on closing position

The impact energy e calculated using equation (5.3) at a closing velocity of 1860 mm/s would be 24 times higher than that at a closing velocity of 375 mm/s – the valve closing velocity at the maximum valve clearance (see Fig. 5.17).

$$e = \frac{1}{2}mv^2 \tag{5.3}$$

where m is the mass of the valve added to the mass of the follower and half the valve spring mass (kg).

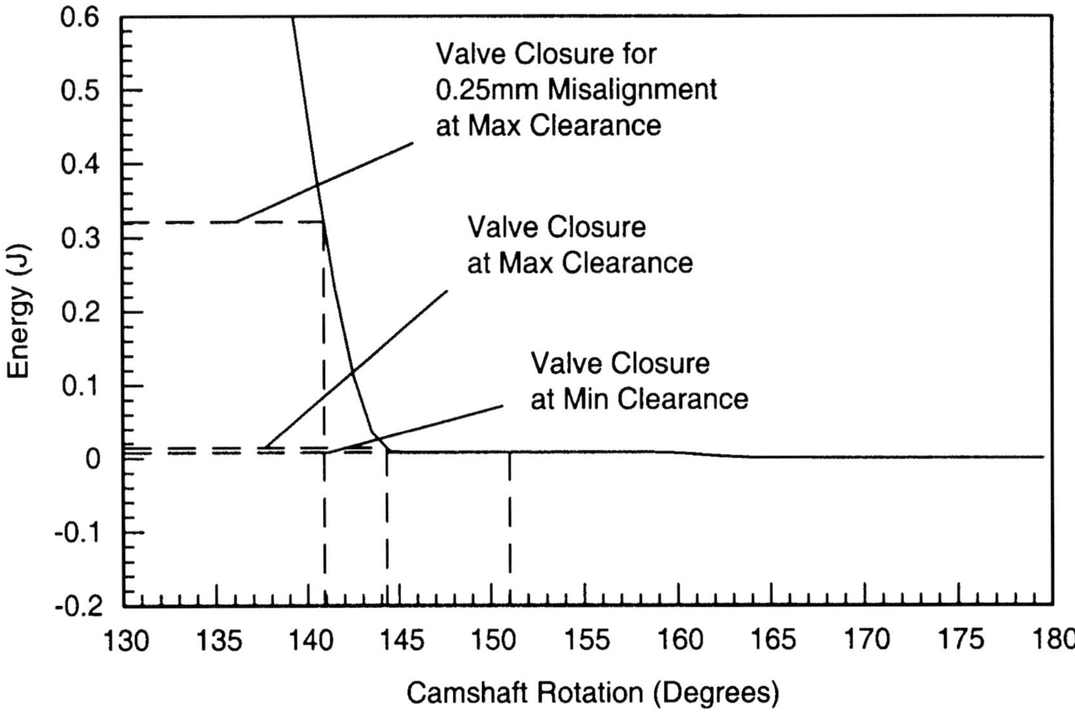

Fig. 5.17 1.8 l, IDI, diesel engine inlet valve energy at valve closure

Work done is equal to force multiplied by distance. In order to calculate the work done on a valve during combustion, W_v, the distance was taken as the maximum valve head deflection, y_{max}, and the force the maximum combustion load, P_p, see equation (5.4)

$$W_v = y_{max} \times P_p \tag{5.4}$$

Assuming that a valve head is a flat circular plate with radius R_v and thickness b (as shown in Fig. 5.18), the maximum valve head deflection was calculated using the equations for the deflection of a simply supported flat circular plate with a uniformly distributed load, see Figure 5.19 and equation (5.5) as outlined by Roark and Young [11]

$$y_{max} = \frac{-P_p R_v^4 (5+v)}{64D(1+v)}, \text{ where } D = \frac{Eb^3}{12(1-v^2)} \tag{5.5}$$

$$\Rightarrow y_{max} = \frac{-3P_p R_v^4 (1-v^2)(5+v)}{16Eb^3(1+v)}$$

$$\Rightarrow y_{max} = \frac{-3P_p R_v^4 (1-v)(5+v)}{16Eb^3} \tag{5.6}$$

where b is the plate thickness (m), v is Poisson's ratio, and E is the modulus of elasticity (N/m^2).

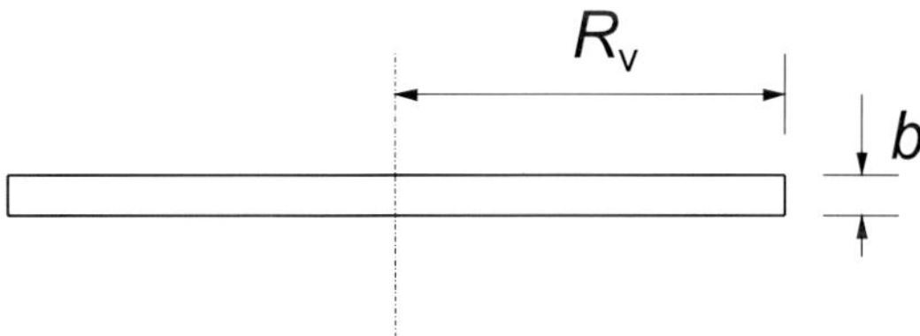

Fig. 5.18 Flat circular plate

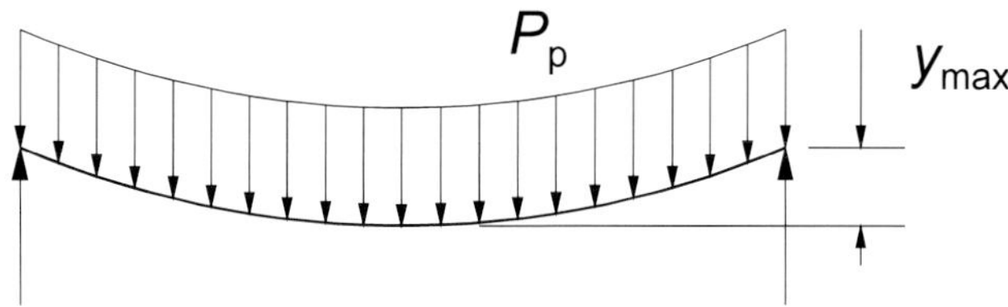

Fig. 5.19 Simply supported circular flat plate with a uniformly distributed load

For a N/A, 1.8 litre, IDI, diesel engine: p_p = 13×10^6 N/m^2; R_v = 18×10^{-3} m; b = 8×10^{-3} m (estimated value as valve head thickness varies); ν = 0.3; and E = 210×10^9 N/m^2.

Therefore, from equation (5.6)

maximum valve head deflection, $y_{max} = -8.8\ \mu$m

For a N/A, 1.8 litre, IDI, diesel engine

P_p = 13 200 N

Therefore, from equation (5.4)

work done on a valve during combustion, W_v = 0.12 J per combustion cycle

5.5.3.2 Hydraulic test machine

The actuator on the hydraulic test machine acts to move the test rig up to and away from the valve which is held in a 'fixed' lateral position by a spring. Calculations were, therefore, based on the actuator motion.

The sinusoidal lift curve for an 'impact and sliding' test (see Section 5.5.1.2) using baseline parameters (see Table 5.5) was determined using the amplitude *a*, frequency *f*, and initial actuator displacement *L*, see equation (5.7). The velocity curve was then determined by differentiating the lift function, see equation (5.8). Both curves are shown in Fig. 5.20.

$$l_a = \alpha \sin(2\pi ft) + L \tag{5.7}$$

where l_a is the actuator lift (m) and t is time (seconds).

$$v_a = \frac{dl_a}{dt} = 2\pi f\alpha \cos(2\pi ft) \tag{5.8}$$

where v_a is the actuator velocity (m/s).

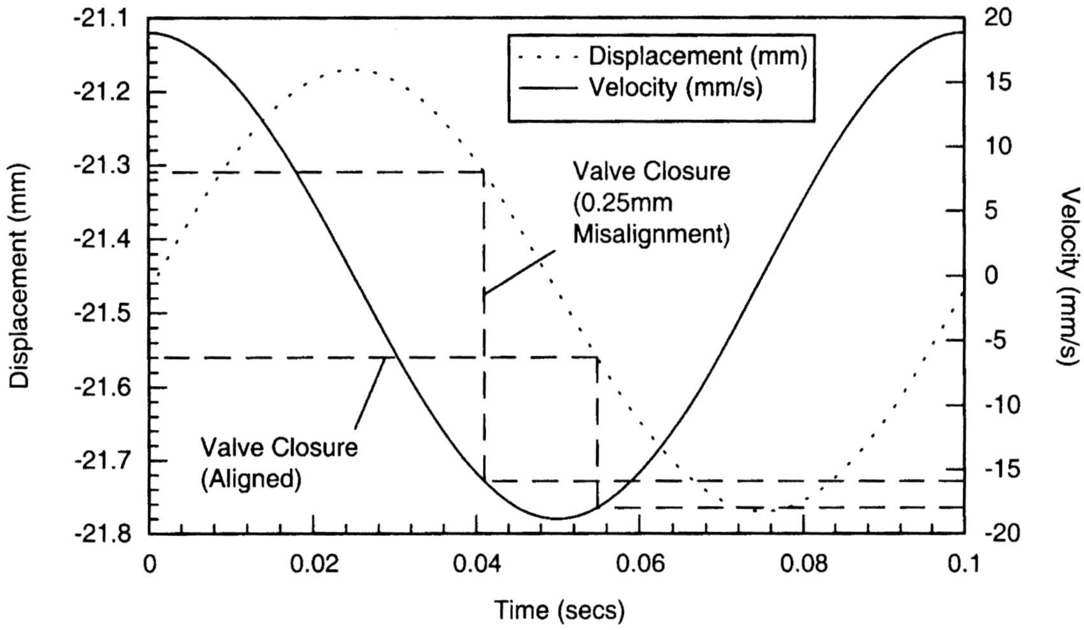

Fig. 5.20 Test rig displacement and velocity in hydraulic test machine (13 kN combustion load)

The velocity at closure for an aligned valve is 18 mm/s. The closing velocity for a valve misaligned by 0.25 mm, also shown in Fig. 5.20, is 16 mm/s.

The energies at valve closure for an aligned and a misaligned valve (shown in Fig. 5.21) are 0.0081 J and 0.0064 J, respectively (calculated using equation (5.3), where m = test rig mass = 50 Kg).

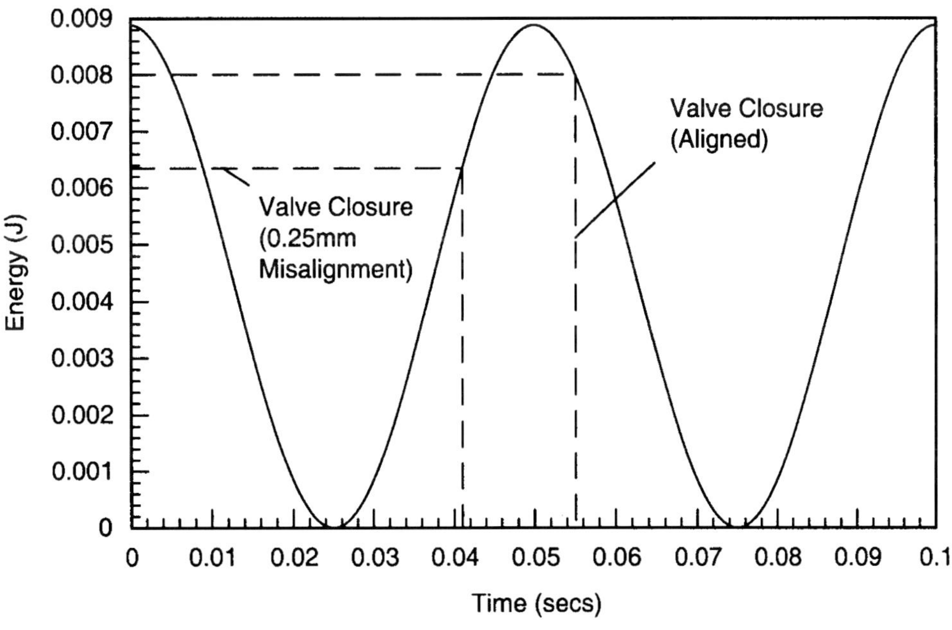

Fig. 5.21 Test rig energy in hydraulic test machine (13 kN combustion load)

In order to calculate the work done on the valve per cycle, the valve head deflection first had to be estimated. The loading data shown in Fig. 5.22 indicate that the total deflection at the baseline 'combustion' loading (13 kN) is 0.15 mm (maximum deflection minus deflection at valve closure; readings taken from the LVDT on the hydraulic test machine).

The deflection of the rig itself at this load was calculated using the equation for the deflection of a built-in flat circular plate with a central load, as outlined by Roark and Young [11], see equation (5.9)

$$y_{\max} = \frac{-P_p R_r^2}{16\pi D}, \text{ where } D = \frac{Eb^3}{12(1-v^2)} \tag{5.9}$$

$$\Rightarrow y_{\max} = \frac{-3P_p R_r^2 (1-v^2)}{4\pi E b^3} \tag{5.10}$$

where $y_{\max}$ is the maximum vertical deflection (m), b is the plate thickness (m), v is Poisson's ratio, E is the modulus of elasticity (N/m²) and R_r is the plate radius (m).

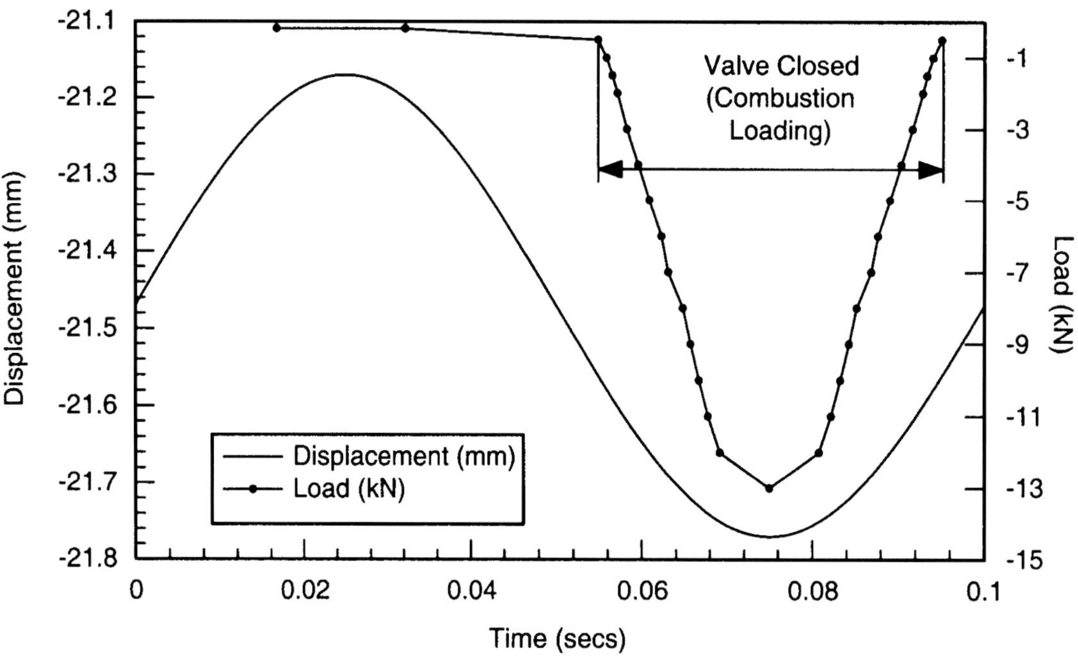

Fig. 5.22 Test rig displacement and actuator load

For the hydraulic loading apparatus at baseline 'combustion' loading: $P_p = 13.2\times10^3$ N; $R_r = 282\times10^{-3}$ m; $b = 20\times10^{-3}$ m; $v = 0.3$; and $E = 210\times10^9$ N/m^2.

(These values were obtained from measurements taken on the rig and from properties of the material used in the manufacture of the rig.)

Therefore, from equation (5.10)

$$y_{max} = -135.7\ \mu\text{m}$$

The valve deflection at the baseline 'combustion' load is equal to the total deflection minus the test rig deflection, therefore

$$\text{valve deflection} = \text{total deflection} - y_{max} = 14.3\ \mu\text{m}$$

Therefore, from equation (5.4)

$$W_v = 0.18 \text{ J per combustion cycle}$$

While the errors apparent in calculating the deflection of the valve and test rig in this manner are potentially high due to the simplifications made, it should be emphasized that the purpose of the calculation was merely to provide a simple comparison between the work done in the test rig with that in the engine rather than provide an accurate assessment of the deflection.

For ease of comparison the valve dynamic and loading characteristics for the 1.8 litre, IDI, diesel engine and the hydraulic test machine are shown in Table 5.6. As can be seen, for baseline conditions (see Table 5.5) the impact energy on valve closure and the work done on the valve during the 'combustion cycle' in the hydraulic loading apparatus are of the same order of those in the engine. The hydraulic loading apparatus is, however, unable to reproduce valve closing velocities that occur in the engine. In the engine an increase in the valve clearance, because of either poor adjustment or misalignment relative to the seat, can massively increase the valve closing velocity and energy. In the hydraulic loading apparatus, however, due to the operating constraints on the hydraulic actuator, such variation cannot be achieved even by introducing misalignment.

Table 5.6 A comparison of valve dynamics and loading in a 1.8 l IDI diesel engine and a hydraulic test machine

	1.8 litre IDI diesel engine			*Hydraulic test machine*	
	Min. clearance	Max. clearance		13 kN combustion load	
	Aligned	Aligned	0.25mm misalignment	Aligned	0.25 mm misalignment
Closing velocity (mm/s)	288	375	1860	18	16
Closing impact energy (J)	0.0076	0.013	0.32	0.0081	0.0064
Work done during combustion (J)		0.12			0.91

It is clear that the hydraulic loading apparatus is able to simulate combustion loading of the valve well. As already observed, however, it is not possible to accurately reproduce the dynamics of valve closure seen in an engine over a range of closing velocities.

Performance charts intended to simplify parameter selection were produced for the hydraulic loading apparatus at the baseline combustion load (13 kN) (as shown in Figs 5.23 and 5.24, respectively).

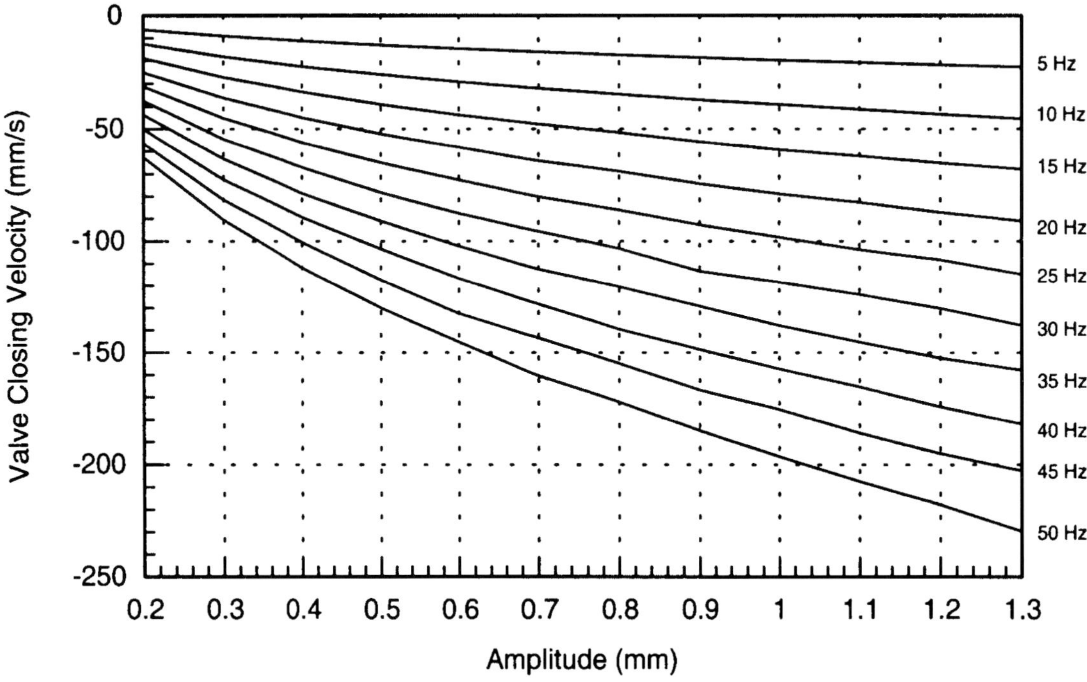

Fig. 5.23 Valve closing velocity against amplitude for varying frequencies (13 kN combustion load)

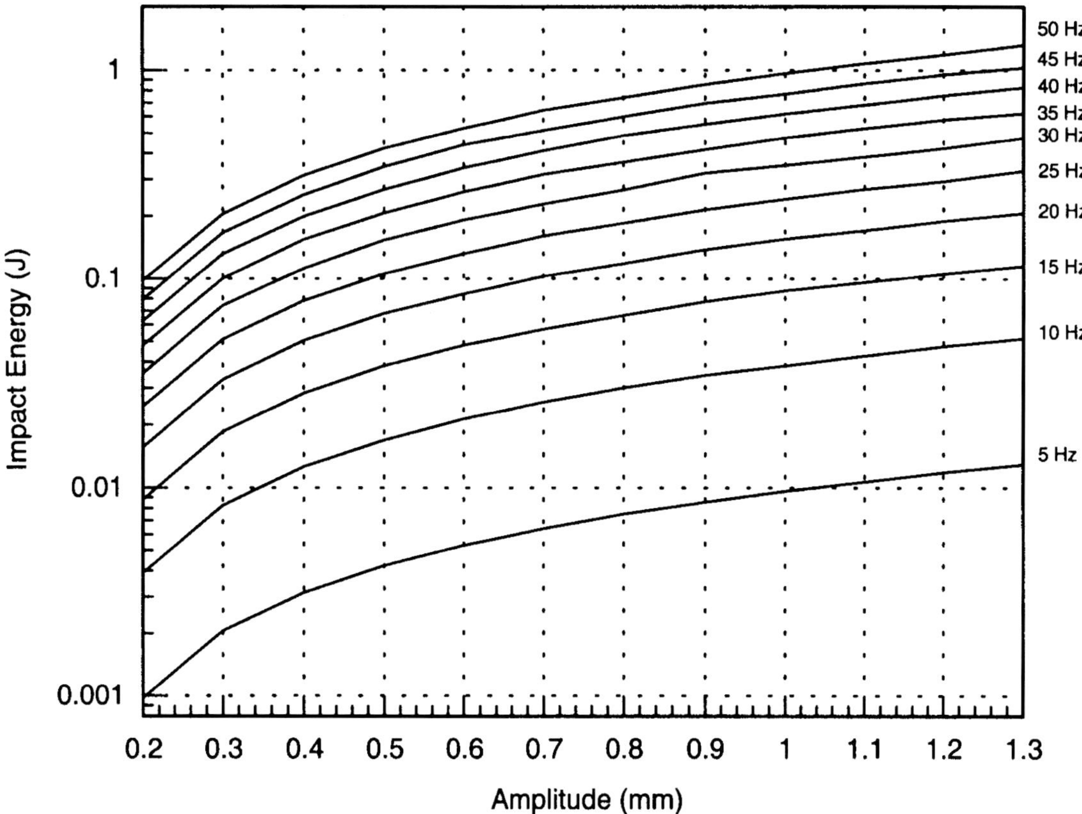

Fig. 5.24 Impact energy on valve closure against amplitude for varying frequencies (13 kN combustion load)

Sinusoidal lift curves for a range of amplitudes and frequencies were differentiated to calculate the corresponding velocity curves, see equations (5.7) and (5.8). The closing velocities taken from these curves were then used to calculate impact energies at valve closure, see equation (5.3). Both closing velocity and impact energy were then plotted against amplitude for a range of frequencies.

5.6 References

1. **Narasimhan, S.L.** and **Larson, J.M.** (1985) Valve gear wear and materials, SAE Paper 851497, *SAE Trans.*, **94**.

2. **Zinner, K.** (1963) Investigations concerning wear of inlet valve seats in diesel engines, ASME Paper 63-OGP-1.

3. **Pope, J.** (1967) Techniques used in achieving a high specific airflow for high-output medium-speed diesel engines, *Trans. ASME, J. Engng Power*, **89**, 265–275.

4. **Matsushima, N.** (1987) Powder metal seat inserts, *Nainen Kikan*, **26**, 52–57, in Japanese.

5. **Blau, P.J.** (1993) Retrospective survey of the use of laboratory tests to simulate internal combustion engine materials tribology problems, ASTM STP Paper 1199.

6. **Hofmann, C.M., Jones, D.R.,** and **Neumann, W.** (1986) High temperature wear properties of seat insert alloys, SAE Paper 860150, *SAE Trans*., **95**.

7. **Nakagawa, M., Ohishi, S., Andoh, K., Miyazaki, S., Mori, K.,** and **Machida, Y.** (1989) Development of hardsurfacing nickel-based alloy for internal combustion engine intake valves, *JSAE Rev*., 68–71.

8. **Fujiki, F.** and **Makoto, K.** (1992) New PM seat insert materials for high performance engines, SAE Paper 920570.

9. **Malatesta, M.J., Barber, G.C., Larson, J.M.,** and **Narasimhan, S.L.** (1993) Development of a laboratory bench test to simulate seat wear of engine poppet valves, *Tribol. Trans*., **36**, 627–632.

10. **Stone, R.** (1992) *Introduction to internal combustion engines*, Macmillan, Basingstoke.

11. **Roark, R.J.** and **Young, W.C.** (1975) *Formulas for stress and strain*, Fifth edition, McGraw-Hill, New York.

Chapter 6

Experimental Studies on Valve Wear

6.1 Introduction

This chapter details bench test work designed to investigate valve and seat wear and isolate critical parameters, the data from which could be used to develop a model to predict valve recession.

The aims of the bench test work were to:

- investigate the wear mechanisms occurring in passenger car diesel engine inlet valves and seat inserts;
- study the effect of engine operating conditions on wear;
- quantify the effect of lubrication at the valve/seat insert contact;
- test potential new seat insert materials and compare the results with those for existing materials.

Testing was carried out using bench test apparatus designed to simulate the loading environment and contact conditions to which the valve and seat insert are subjected (as described in Chapter 5).

6.2 Investigation of wear mechanisms

6.2.1 Experimental details

6.2.1.1 Specimen details

Valve and seat insert materials characteristic of those currently in use in passenger car diesel engine applications were selected for use in the tests, details of which are shown in Table 6.1. The geometries of the valves and seat inserts used are shown in Fig. 5.4. The 'hardness' of selected materials is shown in Table 6.2.

Table 6.1 Valve and seat insert materials

Valve material	Description
V1	Martensitic, low-alloy steel
V2	Austenitic stainless steel
Seat insert material	**Description**
S1 and S2	Sintered martensitic tool steel matrix with evenly distributed intra-granular spheroidal alloy carbides. Metallic sulphides are distributed throughout at original particle boundaries. The interconnected porosity in both is substantially filled with copper alloy throughout.
S3	Cast, tempered martensitic tool steel matrix with a network of carbides uniformly distributed.

Table 6.2 Hardness of valve and seat insert materials

Material designation	Hardness (Hv)
V1 (Valve)	630
S2 (Seat insert)	490
S3 (Seat insert)	490

6.2.1.2 Test methodologies

In order to investigate the effect of the frictional sliding of the valve on the seat insert under the action of the combustion pressure and the effect on wear of combining this with the impact of the valve on the seat insert during valve closure, the two test methodologies developed during initial testing of the hydraulic loading apparatus were employed (see Section 5.5.1.2). These were as follows.

Frictional sliding

The first methodology employed a triangular loading waveform to investigate the effect of combustion loading on valve and seat insert wear. The intention was to isolate the frictional sliding between the valve seating face and the seat insert, hence the valve seating face was in constant contact with the seat insert. Test parameters used for selected tests are shown in Table 6.3. These were based on the baseline established during initial testing of the hydraulic loading apparatus (see Section 5.5.1.3).

Impact and sliding

The second test methodology used a sinusoidal displacement waveform, with valve lifts of up to 1.2 mm and similar peak loads to those used in the frictional sliding tests, to investigate the effect of the impact of the valve on the seat insert as the valve closes, in combination with the combustion loading. Test parameters used for selected tests are shown in Table 6.4. Again, these were based on the baseline established during initial testing of the rig (see Section 5.5.1.3). When parameters such as combustion load were varied, the rig performance charts (see Figs 5.23 and 5.24) were used to select suitable frequencies and amplitudes in order to maintain a constant closing velocity.

Table 6.3 Frictional sliding test parameters

Seat insert material	Valve temp. (°C)	Freq. (Hz)	Load (kN)	Load waveform	Misalignment (mm)	Rotation (r/min)	No. of cycles
S1	R.T.	20	13	Triangular	0	0	500 381
S1	R.T.	20	13	Triangular	0.5	0	506 521
S1	130	20	13	Triangular	0	0	500 016
S3	R.T.	20	13	Triangular	0	0	500 014
S3	R.T.	20	13	Triangular	0.5	0	500 020
S1	R.T.	5	0.6	Sinusoidal	0	1	66 540

Table 6.4 Impact and sliding test parameters

Seat insert material	Valve temp. (°C)	Freq. (Hz)	Valve lift (mm)	Valve closing velocity (mm/s)	Load (kN)	Displacement (waveform)	Misalignment (mm)	No. of cycles	Lubn. (Y/N)
S3	R.T.	10	0.6	18	13	Sinusoidal	0	25 006	N
S3	R.T.	10	0.6	18	13	Sinusoidal	0.25	18 067	N
S2	R.T.	10	0.6	18	13	Sinusoidal	0	39 997	N
S2	R.T.	10	0.6	18	13	Sinusoidal	0.25	24 009	N
S2	130	10	0.6	18	13	Sinusoidal	0	24 371	N
S3	130	10	0.6	18	13	Sinusoidal	0	24 011	N
S3	R.T.	12	0.6	18	6	Sinusoidal	0.25	24 039	N
S3	R.T.	10	0.6	18	18.5	Sinusoidal	0	100 027	N
S2	R.T.	20	1.2	59	13	Sinusoidal	0	24 041	N
S3	R.T.	10	0.6	18	13	Sinusoidal	0	100 179	N
S2	R.T.	10	0.6	18	13	Sinusoidal	0	100 038	N
S3	R.T.	10	0.6	18	13	Sinusoidal	0	100 000	Y

In order to investigate the effect of impact of the valve on the seat insert as the valve closes on the motorized cylinder head, tests were run using two different seat insert materials, one cast (S3) and the other sintered (S2). In each test, different valve clearances were used in order to vary the valve closing velocity. Different clearances were achieved by using seat insert holders of varying thickness. Having chosen the desired valve closing velocities, the valve clearances required to achieve such velocities were determined using valve lift and velocity curves (such as shown in Fig. 5.14). The thickness of the seat insert holders could then be calculated from these clearances. Details of the clearances used and the closing velocities, energies, and forces at each clearance are shown in Table 6.5. Each test was run for 160 000 cycles (chosen in order to fit the test matrix into the available time). Valve rotation was measured for each valve during testing. This was achieved by marking the valve head and then timing a set number of rotations.

Table 6.5 Motorized cylinder head test parameters

Valve clearance (mm)	Closing velocity (mm/s)	Closing energy (J)
0.215	324	0.0096
0.415	960	0.0845
0.515	1600	0.234
0.615	2100	0.404
1.420	3680	1.239

6.2.1.3 Wear evaluation

Wear evaluation was achieved using optical microscopy to study wear scars. Wear features were studied to establish wear mechanisms occurring in the valves and seat inserts. In addition, wear scar widths were measured both during and after the tests.

6.2.2 Results

6.2.2.1 Appearance of worn surfaces

When employing the *frictional sliding* test methodology on the hydraulic loading apparatus to simulate combustion loading, the wear scars achieved on the valve seating faces appeared uneven. The formation of a brown oxide, characteristic of fretting wear (reciprocating sliding wear caused by very small displacements), was observed as well as debris at the edges of the wear scars. The wear scars on valves run against S2 sintered seat inserts were similar to those run against S3 cast seat inserts.

Observation of the seating faces of both the sintered and cast seat inserts revealed the presence of scratches in the radial direction (see Fig. 6.1). These were similar to those previously observed [1].

The unevenness of the wear scars was caused by non-uniform contact between the valve and seat insert. The observations characteristic of fretting and sliding wear verified that the frictional sliding caused by the combustion load had been isolated. The consistency of these observations on both cast and sintered seat inserts indicates that the two materials have a similar resistance to sliding wear.

When employing the *impact and sliding* test methodology on the hydraulic loading apparatus, and thus allowing the valve to lift from the seat during a cycle, the wear scars achieved on the seating face of the valves were more even. Examination of the wear scars of valves run against S3 cast seat inserts revealed evidence of deformation and the presence of a series of ridges and valleys formed circumferentially around the axis of the valve seating face (as shown in Fig. 6.2). These correspond to the 'single wave formation' previously described [1].

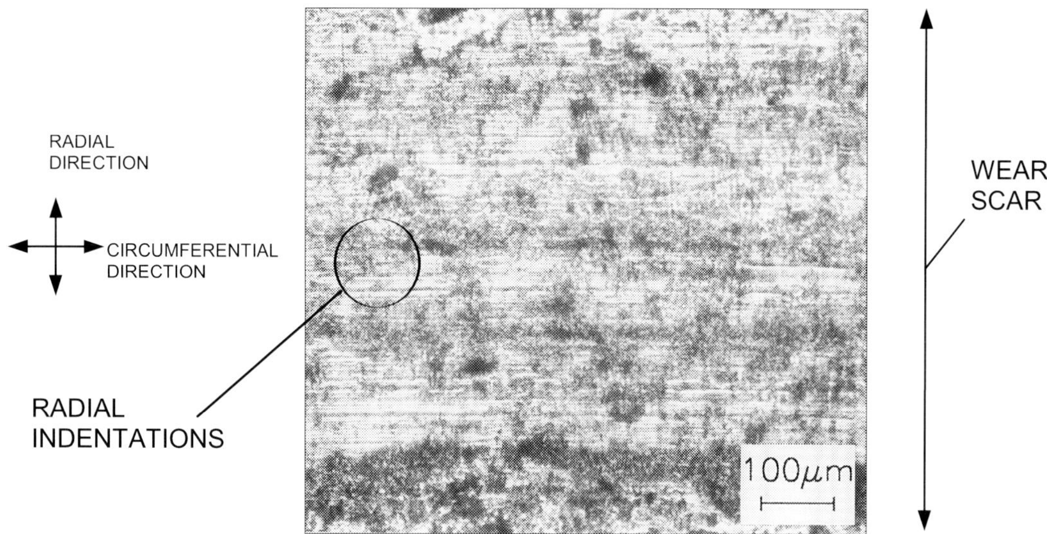

Fig. 6.1 Seat insert seating face showing indentations in the radial direction (valve material V1 run against sintered seat insert material S1 on hydraulic loading apparatus). Major seat diameter is towards top of figure. See Table 6.1 for details of materials. (Reprinted with permission from SAE paper 1999-01-1216 © 1999 Society of Automotive Engineers, Inc.)

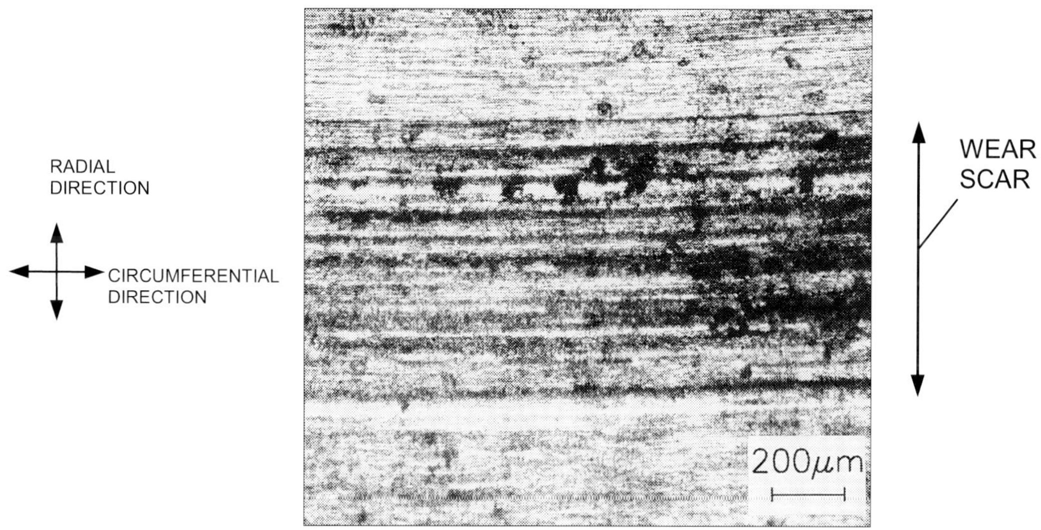

Fig. 6.2 Valve seating face showing a series of ridges and valleys formed circumferentially around the axis of the valve (valve material V1 run against cast seat insert material S3 on hydraulic loading apparatus). Major seat diameter is towards bottom of figure. See Table 6.1 for details of materials. (Reprinted with permission from SAE paper 1999-01-1216 © 1999 Society of Automotive Engineers, Inc.)

This type of deformation was less prevalent on valves run against S2 sintered seat inserts. Evidence was found, however, of adhesive pick-up from seat inserts (see Figs 6.3 and 6.4). With a cast insert there appeared to be greater surface damage to the valve than the seat insert, whereas with a sintered seat insert there appeared to be greater insert damage.

Fig. 6.3 Valve seating face showing evidence of adhesive pick-up from the seat insert (valve material V1 run against sintered seat insert material S2 on hydraulic loading apparatus). Major seat diameter is towards bottom of figure. See Table 6.1 for details of materials

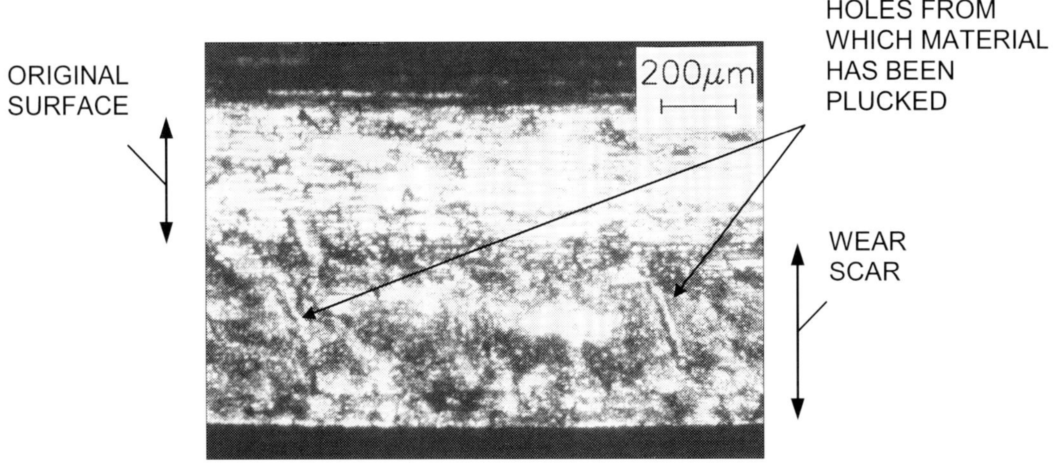

Fig. 6.4 Seat insert seating face showing evidence of adhesive pick-up (valve material V1 run against sintered seat insert material S2 on hydraulic loading apparatus). Major seat diameter is towards top of figure. See Table 6.1 for details of materials

Observation of the seating faces of both types of seat insert revealed that surface damage was more severe when combining impact with frictional sliding. A wear scar was seen to form and grow on sintered seat inserts, whereas on the cast seat inserts no obvious scar formed, although evidence was found of pitting and radial indentations.

The results achieved using the two different test methodologies, *frictional sliding* and *impact and sliding*, indicate that the effect of the two loads imposed on the valve have been isolated. The *frictional sliding* test was used to simulate the load imposed during combustion in the cylinder and the *impact and sliding* test was used to simulate the impact load on valve closure in combination with the combustion loading. The radial scratches on the seat inserts found when using the loading waveform are caused as the valve slides against the seat insert as it deflects under loading. The circumferential ridges and valleys found when using the displacement waveform are caused by a deformation or gouging process as the valve impacts against the seat insert.

Both impact and sliding clearly have a large influence on valve recession. It is in combination, however, that they have the largest effect. In the tests with impact and sliding run in combination, it took a few thousand cycles to achieve surface damage attained in several hundred thousand cycles in the frictional sliding tests.

Tests run on the motorized cylinder head (without any combustion loading) were intended to isolate the impact of the valve on the seat insert on valve closure. Evidence was found on valves run with a high closing velocity, however, that a small amount of sliding was occurring even in the absence of the combustion loading. It was also found on removing the valves, that a film of oil was present on the valve head and seating face. This was formed by lubricating oil leaking past the valve stem seals.

As with tests run on the hydraulic loading apparatus using impact and sliding, valve wear was observed to be far more severe with S3 cast seat inserts, and insert wear was more severe with S2 sintered seat inserts.

Examination of the wear scars of valves run against S3 cast seat inserts again revealed evidence of deformation. There was also a series of ridges and valleys formed circumferentially around the axis of the valve seating face, similar to those observed during testing on the hydraulic loading apparatus. Observation of the cast seat insert seating faces (see Fig. 6.5), however, revealed the presence of surface cracking and evidence of subsequent material loss, not previously observed.

The wear features observed on both valves and seat inserts (deformation and surface cracking) are characteristic of processes resulting in wear loss due to single or multiple impact of particles [2] (see Fig. 6.6). Similar observations made during work on the wear of poppet valves operating in hydropowered stoping mining equipment led Fricke and Allen [3] to use a relationship of the same form as that used in erosion studies to model impact wear. Fricke and Allen [3] justified the use of such a relationship for impact wear of valves, citing work by Hutchings *et al.* [4] in which it was shown that erosion can be satisfactorily modelled by the impact of large particles. In their work, they used hard steel balls up to 9.5 mm in diameter. It was thus assumed that a relationship exists between impacts on a macroscale (greater than 1 mm) and impacts on a microscale (less than 1mm), such as those found typically in erosive wear.

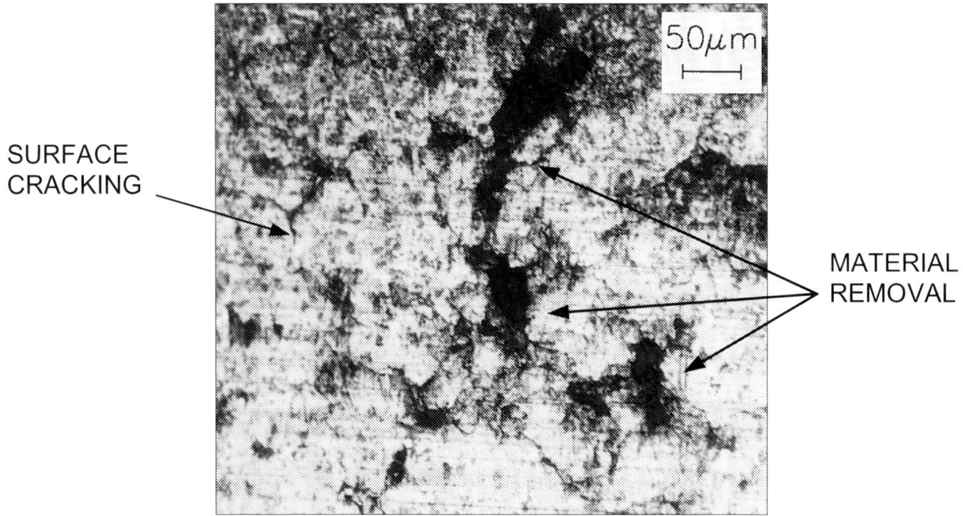

Fig. 6.5 Seat insert seating face (valve material V1 run against cast seat insert material S3 on motorized cylinder head). Major seat diameter is towards top of figure. See Table 6.1 for details of materials

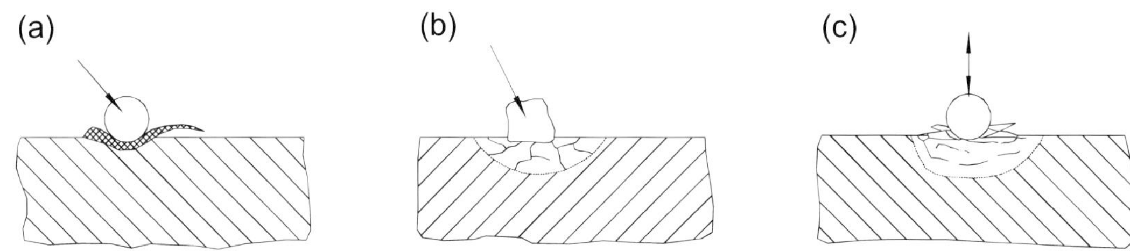

Fig. 6.6 Processes resulting in wear loss due to single or multiple impact of particles: (a) extrusion of material at the exit end of impact craters; (b) surface cracking (microcracking); (c) surface and subsurface fatigue cracks due to repeated impact [2]

6.2.2.2 Formation of wear scars

Unlike the valve wear scars observed when applying the *frictional sliding* test methodology on the hydraulic loading apparatus, those seen when employing the *impact and sliding* test methodology were uniform and seen to increase in width as the tests proceeded (it should be noted that during these tests no valve rotation was used). The progression of wear when using a sintered seat insert differed from that when using a cast seat insert (as shown in Fig. 6.7).

The observations made indicate that the introduction of impact caused a bedding-in process to occur, improving the uniformity of the valve wear scars and causing them to increase in width. Figure 6.7 shows that, initially, there was a rapid increase in the wear scar width; this progression then slowed until bedding-in was achieved. It also shows how the cast and sintered seat inserts responded differently to the introduction of impact. When using a cast insert, a high initial progression of wear was observed compared to the more gradual progression when using a sintered seat insert.

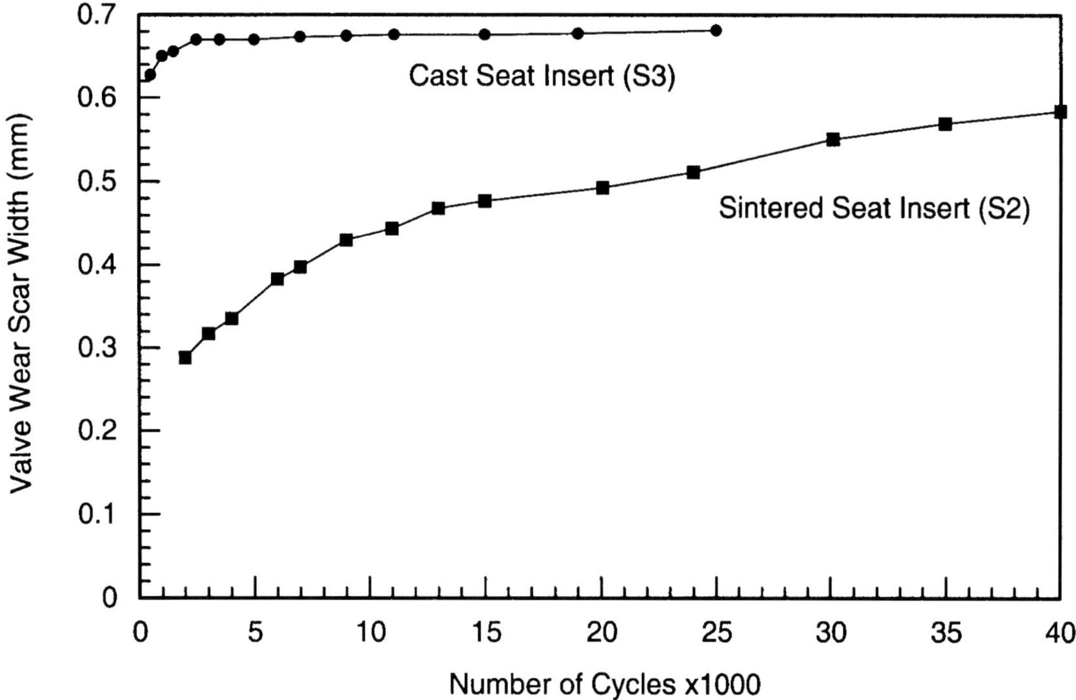

Fig. 6.7 Average wear scar width for a V1 valve run against an S2 sintered seat insert and a V1 valve run against an S3 cast seat insert using impact and sliding on the hydraulic loading apparatus. See Table 6.1 for details of materials

Recession is the main parameter of interest to engine developers. Wear scar measurement, while providing information on the progression of wear during a test, gave no indication of the actual magnitude of the wear or valve recession that had occurred as no account could be taken of initial contact conditions. In order to obtain a more useful indication of wear, give an improved comparison of test rig data, and provide a means to compare test rig data with valve recession data from engine tests, a method was developed for estimating a recession value from seat insert wear scar data.

Two different wear 'cases' were observed during testing.

1. The valve and seat insert seating face angles differed slightly and a wear scar was seen to form and grow on the valve and seat insert until full contact with the seat insert seating face was achieved, the width of which then began to increase as the test progressed.
2. The valve initially made full contact with the seat insert seating face, which was then seen to grow as the test progressed.

Figure 6.8 shows a comparison of valve recession for *impact and sliding* tests run with cast and sintered seat inserts on the hydraulic loading apparatus (replot of data from Fig. 6.7), calculated using equations that give recession and wear volume as a function of the seat insert wear scar width or seating face width [5]. This shows that valve recession is higher when using sintered seat inserts despite the lower wear scar widths observed.

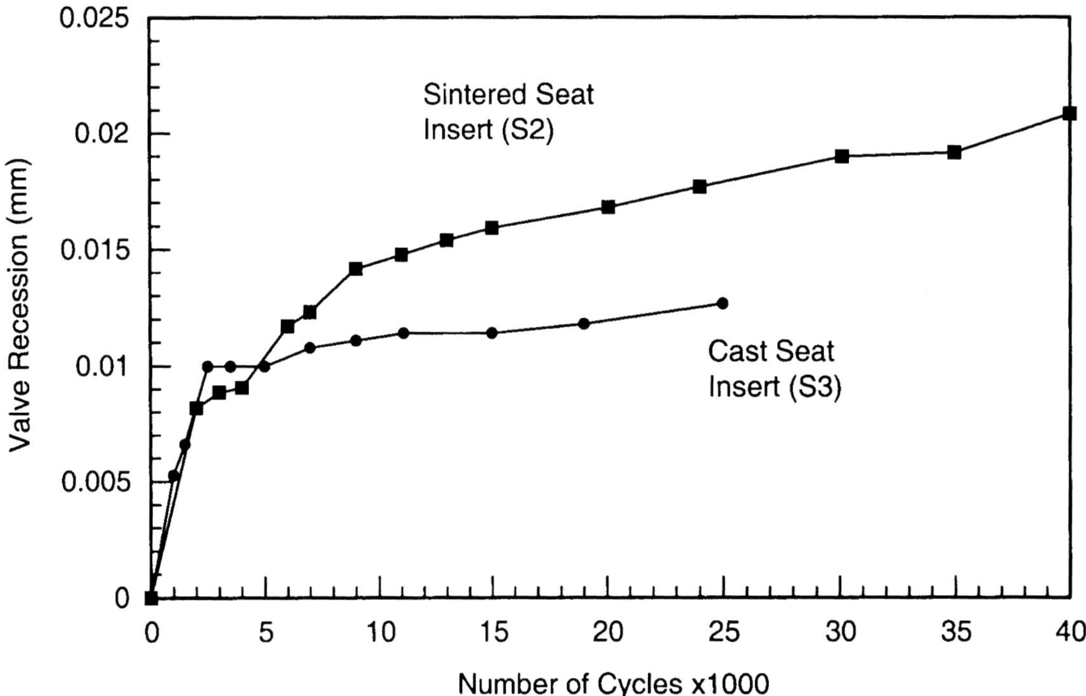

Fig. 6.8 Valve recession for a V1 valve run against an S2 sintered seat insert and a V1 valve run against an S3 cast seat insert using impact and sliding on the hydraulic loading apparatus. See Table 6.1 for details of materials

Figure 6.9 shows a comparison of valve recession (calculated from wear scar and seating face width data) for tests run with cast and sintered seat inserts on the motorized cylinder head. It can be seen, as with the hydraulic loading apparatus results, that greater valve recession occurs when using a sintered seat insert. The tests run on the motorized cylinder head were intended to isolate the impact of the valve on the seat insert on valve closure. These results, therefore, indicate that the cast seat insert material has a greater resistance to impact than the sintered material.

It was clear from the wear scars observed on both valves and seat inserts after tests run on both the hydraulic loading apparatus and the motorized cylinder head, that the wear mechanisms were different for the cast and sintered seat inserts. A sintered seat insert appeared to wear at a greater rate than the valve, i.e. the valve was 'bedding' into the seat insert (see Fig. 6.10). When using a cast seat insert, however, it appeared that the valve was wearing more than the seat insert and that the seat insert was 'bedding' into the valve (see Fig. 6.10).

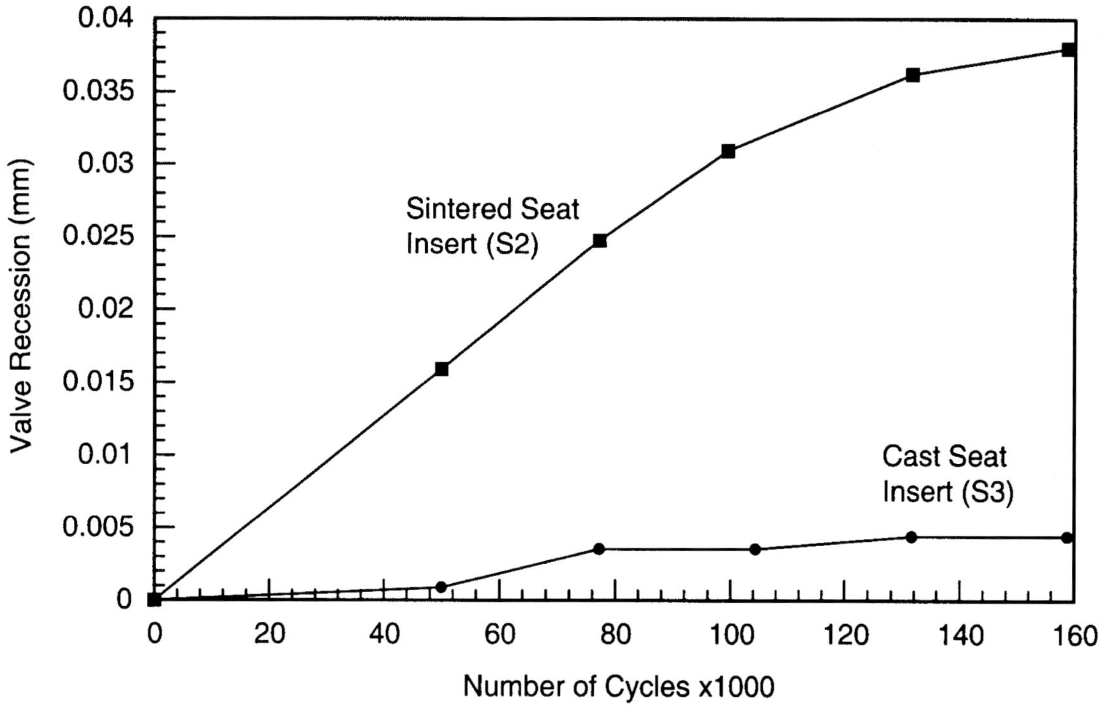

Fig. 6.9 Valve recession for a V1 valve run against an S2 sintered seat insert and a V1 valve run against an S3 cast seat insert with a valve closing velocity of 960 mm/s on the motorized cylinder head. See Table 6.1 for details of materials

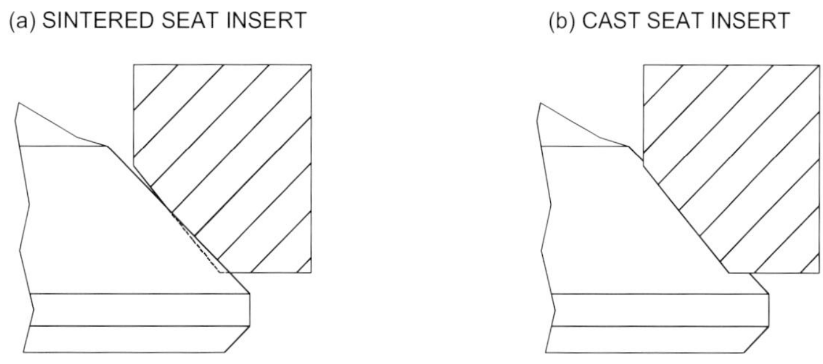

Fig. 6.10 Sintered seat insert versus cast seat insert. (Reprinted with permission from SAE paper 1999-01-1216 © 1999 Society of Automotive Engineers, Inc.)

Both the cast and sintered seat insert materials have similar hardness. It has been shown, however, that with impact wear there is no direct correlation between the material loss and hardness [6]. The fracture toughness of a material has been shown to be one of the important factors controlling impact wear [7]. The cast insert material has a higher fracture toughness than the sintered material and is, therefore, more resistant to the impact load imposed during valve closing, which could explain the two different bedding-in processes observed. It should be noted that the cast and sintered materials did not have the same composition, which could also help to explain the differences in response to impact.

Part of the reason for the development of sintered seat inserts was to allow the incorporation of solid lubricants into the matrix to reduce sliding wear in 'dry' running conditions brought about by the reduction in lead in gasoline. The *frictional sliding* tests have shown that sintered seat insert materials perform adequately under these conditions. However, in order to increase the performance of sintered materials in seat insert applications, the issue of resistance to impact may need to be readdressed.

Material choice is clearly critical in addressing valve and seat insert wear problems. In deciding which material combination to use, consideration should be given to deciding which is more preferable: greater valve wear or greater seat insert wear.

Clearly, replacing valves is less costly than replacing seat inserts or an entire cylinder head. Therefore, work needs to be focussed on reducing seat wear while keeping valve wear at an acceptable level. Ultimately, however, it would be preferable not to have to replace either and to reduce the adjustment required on valve clearances, as a result of recession, to an absolute minimum.

In selecting materials, consideration must be given to the relative resistance required to sliding and impact wear. This can be determined by looking at the engine operating parameters. If a high valve mass or closing velocity is being used then resistance to impact is critical. However, if high peak combustion loads are in use, then resistance to sliding could be more important.

6.2.2.3 Comparison with engine recession data

Hydraulic loading apparatus wear scar width data were used to calculate recession values using the equations relating recession to wear volume [5] (see Section 6.2.2.2). A line was fitted to the recession data using an exponential relationship. This was then extrapolated in order to allow a comparison with engine test data.

Figure 6.11 compares extrapolated hydraulic loading apparatus recession data for a test run with a V1 valve against an S2 sintered seat insert and engine test data for a V1 valve run against an S1 sintered seat insert, alongside hydraulic loading apparatus and engine test data for tests run with a V1 valve against an S3 cast seat insert (hydraulic loading apparatus data points were calculated to correspond with available engine test data points, hence the extrapolated line is not a smooth curve). The engine tests were run at a speed of 4800 r/min.

Good correlation was achieved between the hydraulic loading apparatus and engine test rig data, which further established the validity of the test methodology and indicates the suitability of calculating recession values from wear scar data.

The differences between the hydraulic loading apparatus test operating conditions and those found in an engine, however, should have given rise to higher recession rates in the test rig (hydraulic loading apparatus tests were run unlubricated and service temperatures were not replicated). The fact that this did not occur could have been because of the differences in valve dynamics in the hydraulic loading apparatus and an engine, as explained in Section 5.5.3. It should also be noted that the two sintered materials used differ in composition. S1 contains a solid lubricant while S2 does not.

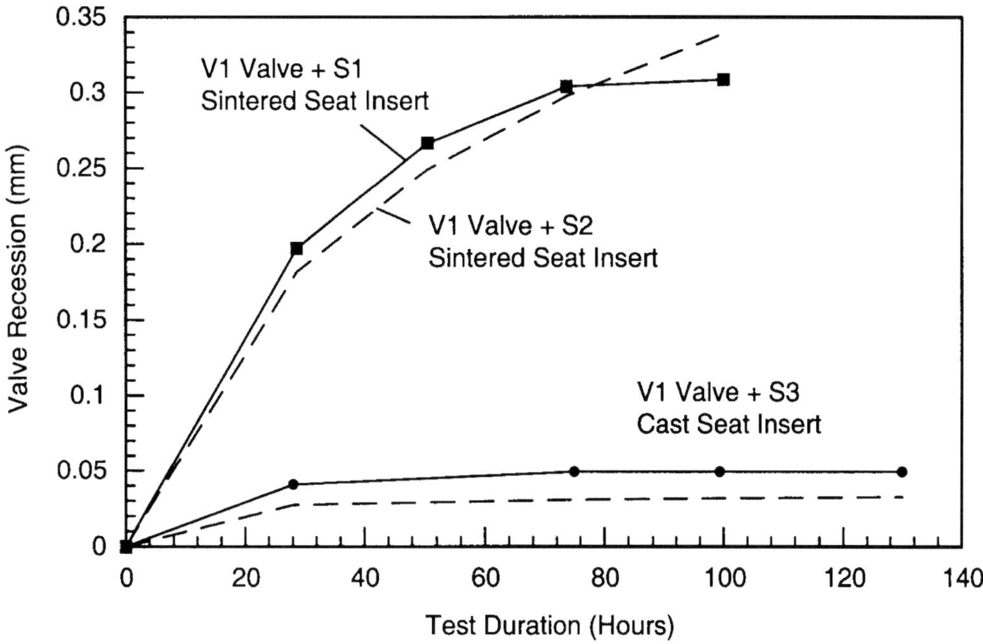

Fig. 6.11 Comparison of hydraulic loading apparatus results with engine test data (solid line – engine test data; broken line – extrapolated test rig data. See Table 6.1 for details of materials

6.2.2.4 Lubrication of valve/seat interface

When running the hydraulic loading apparatus with a steady supply of lubricant to the valve/seat insert interface, evidence of wear on the valve or seat insert seating faces was barely visible. Figure 6.12 compares valve recession (calculated from wear scar data) for tests run with and without lubrication. Although the lubricant was supplied in a larger amount and at a lower temperature than would be experienced in an engine, this provides an approximate quantification of the effect of lubrication on valve and seat insert wear. For this particular case, recession is approximately 3.5 times higher for the test run without lubrication.

It is thought, due to the relatively slow sliding velocity and high load, that the lubrication mechanism occurring during the sliding at the valve/seat interface is that of boundary lubrication. In such a lubrication regime the lubricant film thickness is too small to give full fluid film separation of the surfaces, and surface asperities come into contact.

One of the reasons for the work under discussion is the impending reduction of oil in the air stream of diesel engines which will reduce the amount of lubricant at the inlet valve/seat interface. These results give an indication of the increase in wear likely when this occurs.

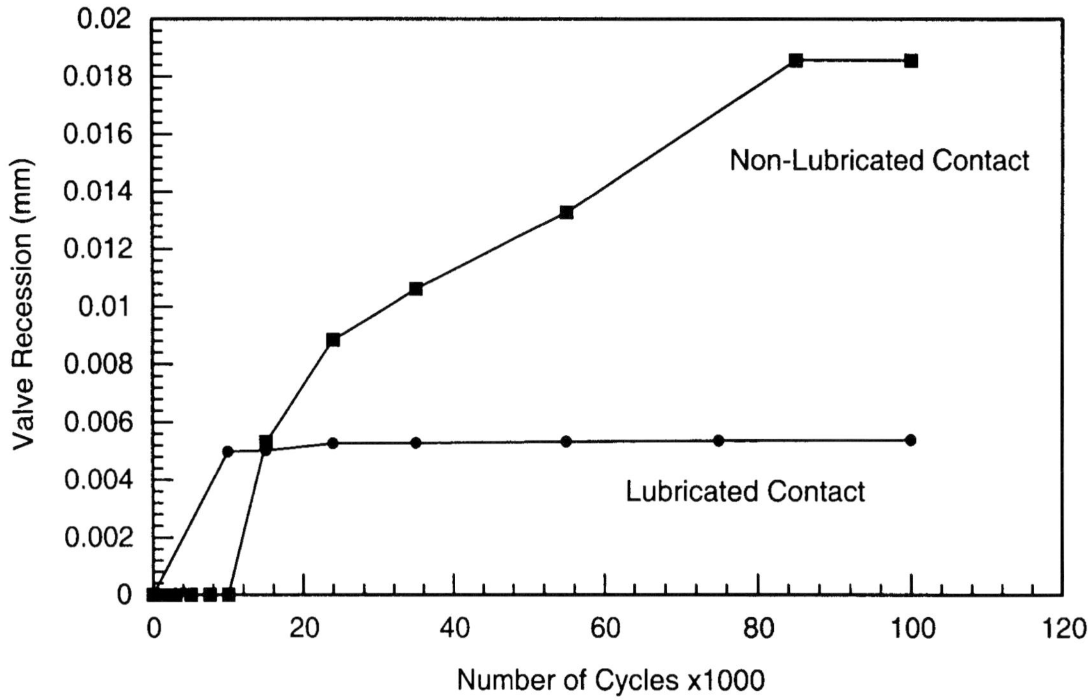

Fig. 6.12 Valve recession for V1 valves run lubricated and unlubricated against an S3 cast seat insert using impact and sliding on the hydraulic loading apparatus. See Table 6.1 for details of materials

6.2.2.5 Misalignment of valve relative to seat

It was found that the magnitude and uniformity of the wear when running valve and seat inserts in an 'aligned' position on the hydraulic loading apparatus were principally affected by the initial contact conditions. Slight misalignments caused by small differences in roundness or seating face angles led to uneven wear. This has also been observed in engine testing. Differences in contact area around a valve seating face can also lead to non-uniform heat transfer from the valve head to the seat insert, causing hot spots or thermal distortions that worsen the problem.

Misaligning the valve relative to the seat insert in the hydraulic loading apparatus produced a crescent-shaped wear scar on the valve seating face (as sketched in Fig. 6.13). The widest point of the scar was at the point of initial contact with the seat insert. At this point, the wear was more severe than was achieved when the valve was aligned. When using *impact and sliding* to simulate impact and combustion loading, the deformation observed on the valve run against a cast insert was more severe at the point of initial contact than when the valve was aligned with the seat insert. The equivalent point on the seat insert also showed evidence of severe wear. Plastic deformation (not observed with an aligned valve) was also found on the valve seating face when misaligning the valve relative to a sintered seat insert (as shown in Fig. 6.14). The comparison of wear scar widths for both aligned and misaligned valves run against S2 sintered seat inserts, shown in Fig. 6.15, gives an indication of the increase in the wear scar width, at the point of initial contact, when misaligning the valve.

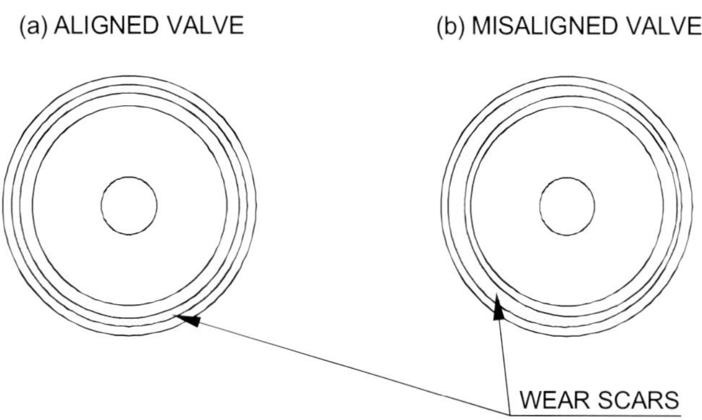

Fig. 6.13 Aligned versus misaligned valve wear scars

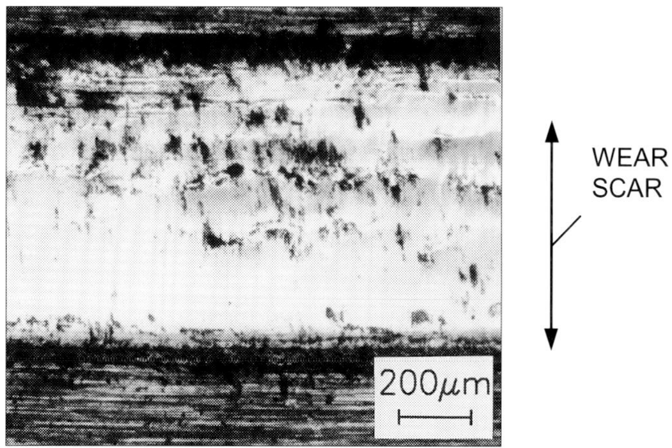

Fig. 6.14 Valve seating face (valve material V1 run misaligned against sintered seat insert material S2 on hydraulic loading apparatus). Major seat diameter is towards bottom of figure. See Table 6.1 for details of materials

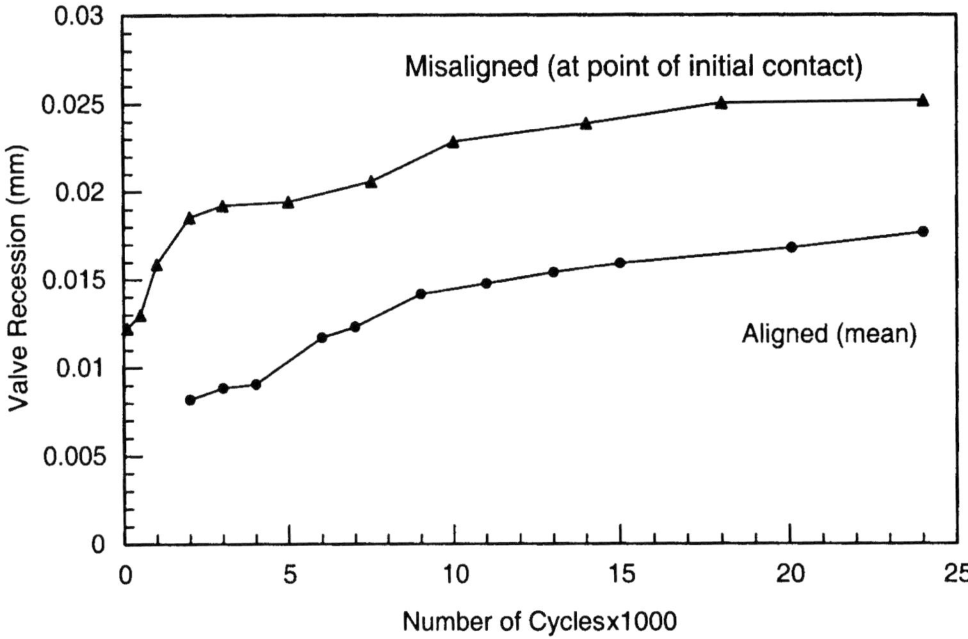

Fig. 6.15 Valve wear scar width for a V1 valve run aligned and misaligned against an S2 sintered seat insert using impact and sliding on the hydraulic loading apparatus. See Table 6.1 for details of materials

The deformation caused by impact on valve closure is increased, as misaligning the valve decreases the initial contact area between the valve and seat insert. The effect of the frictional sliding is increased and, as a result, the wear scar width is increased, since misalignment leads to an increase in the sliding distance as the valve head flexes under the action of the combustion loading.

The magnitude of misalignment was an arbitrary value taken to investigate the effect of valve misalignment (0.25 mm). Actual values have not been measured. Depending, however, on the accuracy of the seat machining and the flexibility of the engine head, it is conceivable that such misalignment may occur in an engine and, therefore, be a source of valve recession.

6.2.2.6 Effect of combustion load

When using the hydraulic loading apparatus, increasing the applied combustion loading while maintaining the closing velocity increased the severity of the wear on both the valves and the seat inserts. It was also observed that at higher loads there were a greater number of radial scratches present on the seat insert seating faces. Figure 6.16 shows recession rates (calculated from wear scar widths) for valves run against cast seat inserts for combustion loads of 6 kN, 13 kN, and 18.5 kN. The increase in wear as the combustion load rises is caused by the increasing effect of frictional sliding as a result of the increase in the combustion loading. The lack of recession at 13 kN for 10 000 cycles was an anomaly. Wear damage was observed during this period, but no measurable wear scar. The recession observed from 10 000 cycles compared well with other tests run at 13 kN.

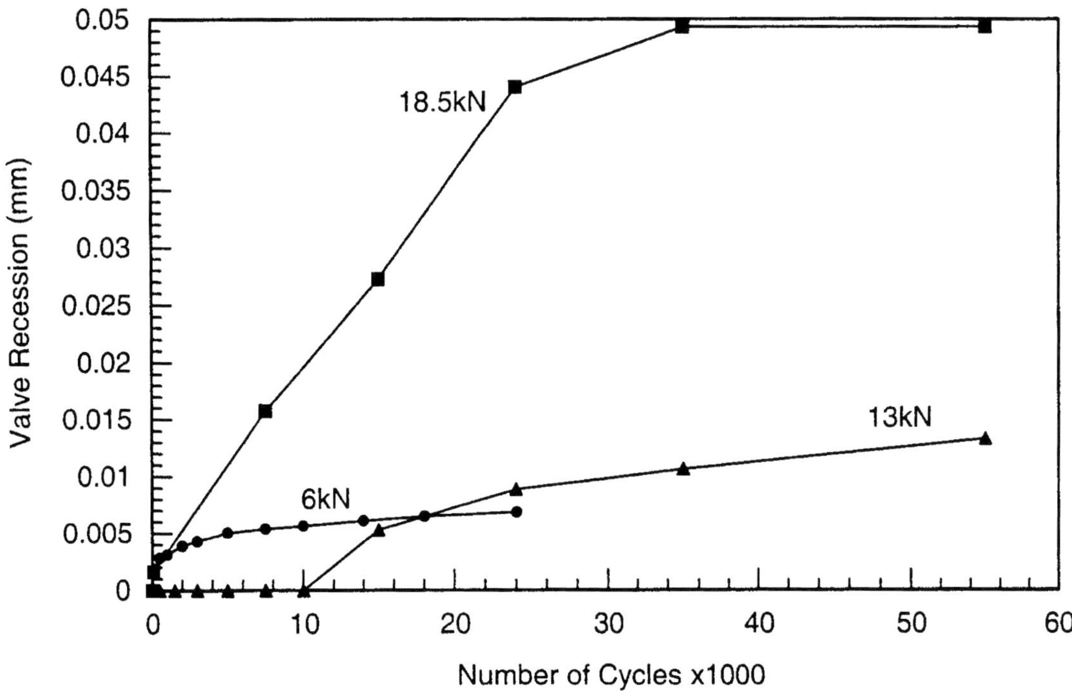

Fig. 6.16 Valve recession for V1 valves run at three different combustion loads against S3 cast seat inserts using impact and sliding on the hydraulic loading apparatus. See Table 6.1 for details of materials

6.2.2.7 Effect of closing velocity

Tests run on the motorized cylinder head – varying the valve closing velocity, clearly indicated that increasing valve closing velocity increased both valve and seat insert wear. As shown in Figs 6.17 and 6.18, valve recession (calculated from wear scar and seating face width data) rapidly increased, when using both cast and sintered insert materials, as the closing velocity was increased.

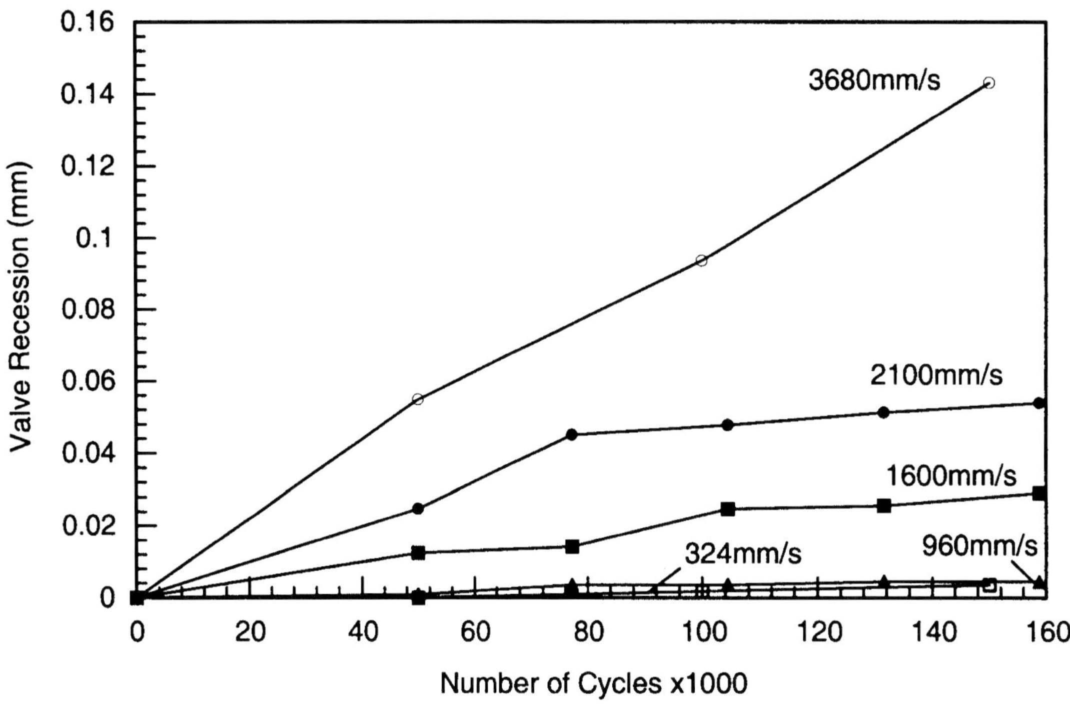

Fig. 6.17 Valve recession for V1 valves run with different closing velocities against S3 cast seat inserts on the motorized cylinder head. See Table 6.1 for details of materials

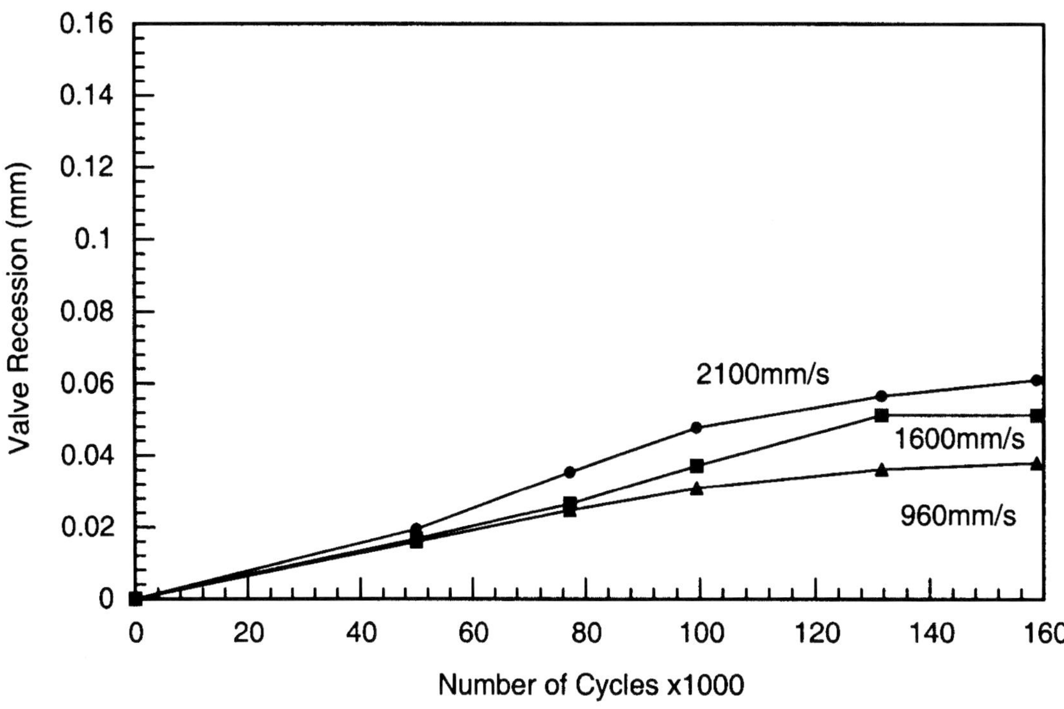

Fig. 6.18 Valve recession for V1 valves run with different closing velocities against S2 sintered seat inserts on the motorized cylinder head. See Table 6.1 for details of materials

Figure 6.19 shows wear scars for valves run against cast seat inserts at three different closing velocities. Deformation and ridge formation was visible as well as fine pitting at higher velocities. The appearance of the wear scar at the lowest velocity shown [Fig 6.19(a)] may have been a due to waviness of the unworn surface. It is interesting, however, that two ridges formed at 1600 mm/s [Fig. 6.19(b)] and only one at 2100 mm/s [Fig 6.19(c)].

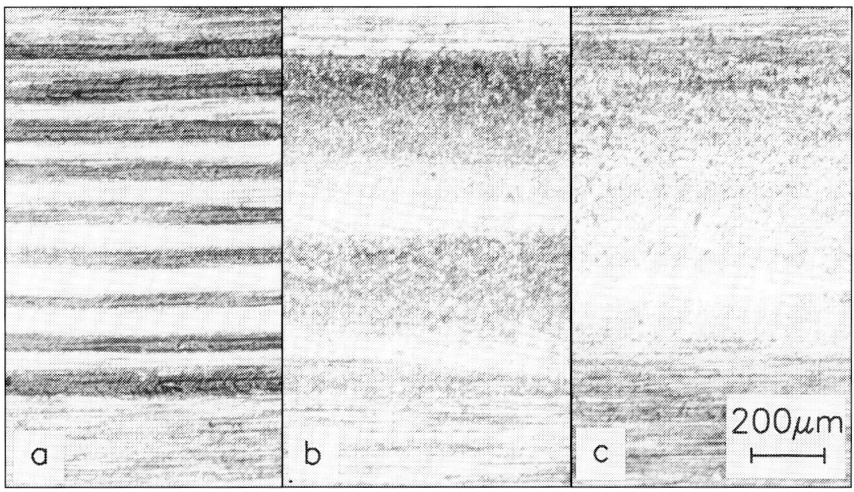

Fig 6.19 Seating faces of valves run with closing velocities of: (a) 960 mm/s; (b) 1600 mm/s; (c) 2100 mm/s. Valve material V1 run against cast seat insert material S3 on motorized cylinder-head. Major seat diameter is towards bottom of figure. See Table 6.1 for details of materials

Figure 6.20 clearly shows that the cast seat insert wear also became more severe as the valve closing velocity was increased. At 960 mm/s [Fig. 6.20(a)] the original machining marks are still visible. As the velocity was increased, however, these became less visible and at 2100 mm/s [Fig. 6.20(c)] they have been completely worn away and surface cracking and subsequent material loss is visible.

Figure 6.21 (a replot of data from Figs 6.17 and 6.18) shows the relationship between recession and valve closing velocity for both sintered and cast seat inserts (at 160 000 cycles). Valve recession is roughly proportional to velocity squared for the cast insert material. The increase in recession observed is similar to that described by Zum-Gahr [2] for the increase in erosion rate with impact velocity of small particles.

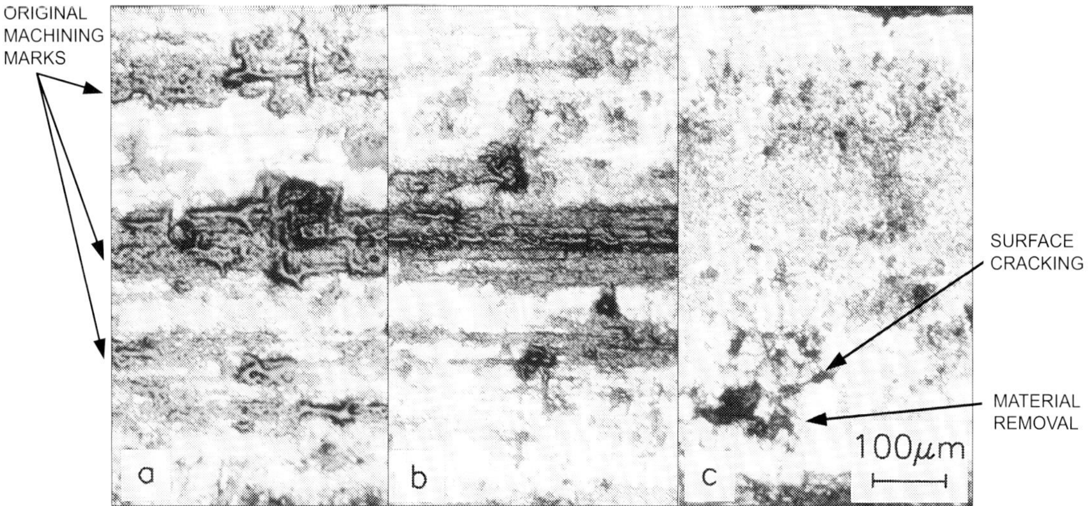

Fig. 6.20 Seating faces of seat inserts run against valves with closing velocities of: (a) 960 mm/s; (b) 1600 mm/s; (c) 2100 mm/s. Valve material V1 run against cast seat insert material S3 on motorized cylinder head. Major seat diameter is towards top of figure. See Table 6.1 for details of materials

6.2.2.8 Valve rotation

When using rotation on the hydraulic loading apparatus, an even wear scar was achieved. Debris was observed at the valve/seat insert interface during testing (as shown in Fig. 6.22). The material was dark and powder-like in nature. It was being removed from the interface under the action of the rotation. It is possible, therefore, that valve rotation promotes debris removal. This would be useful in reducing abrasive wear caused by wear debris otherwise trapped in the interface and in reducing the build-up of deposits and the subsequent formation of hot spots. Examination of the valve seating face of a rotated valve revealed the presence of circumferential grooves. These were caused by rotation of the valve either on closing or while the valve was closed.

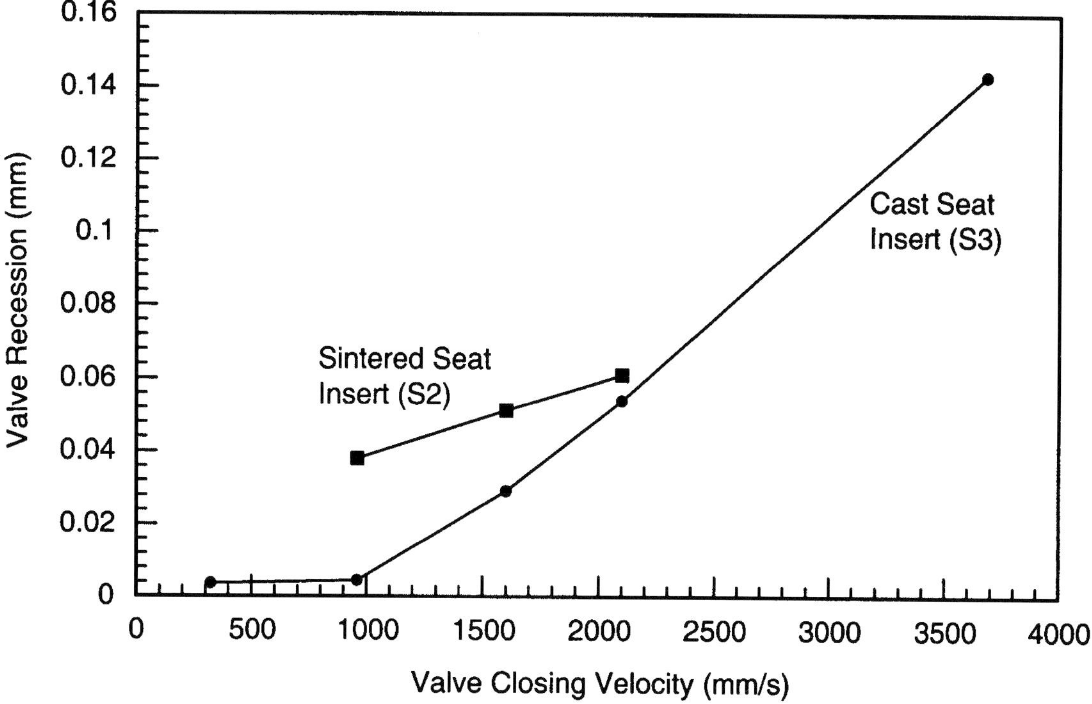

Fig. 6.21 Valve recession against closing velocity for a V1 valve run against an S2 sintered seat insert and a V1 valve run against an S3 cast seat insert on the motorized cylinder head (at 160 000 cycles). See Table 6.1 for details of materials

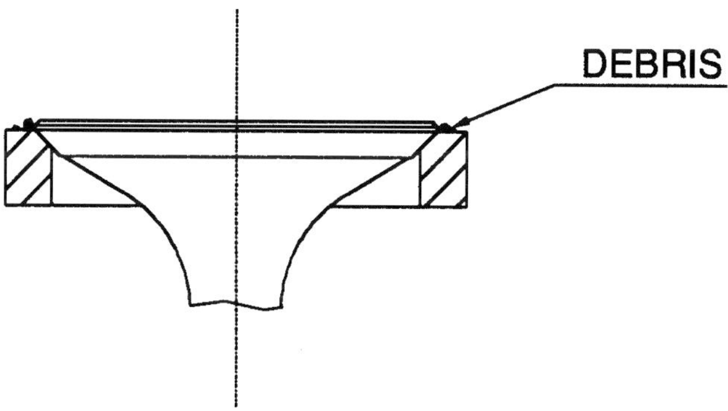

Fig. 6.22 Debris at valve/seat insert interface during valve rotation

Valve rotational speeds were measured during all tests run on the motorized cylinder head. This was achieved by marking the valve heads and then timing a set number of revolutions. Valve rotation was seen to vary between 8 r/min and 18 r/min, depending on test conditions and seat materials.

Valve rotation was observed to decrease with an increase in lubricant supply to the cam/follower interface. This can be explained by looking at the mechanism by which valve rotation occurs. The cam is offset from the follower in order to prevent localized wear at the contact area. The offset cam also promotes valve rotation as a result of the follower rotation, when the valve and follower are in contact. An increase in the lubricant supply to the valve/follower contact will reduce the coefficient of friction. The frictional force will, therefore, also be reduced. This will decrease the rotational speed of the follower, and hence the valve will also rotate at a slower speed. As shown in Fig. 6.23, valve rotation was also observed to decrease with an increase in valve closing velocity.

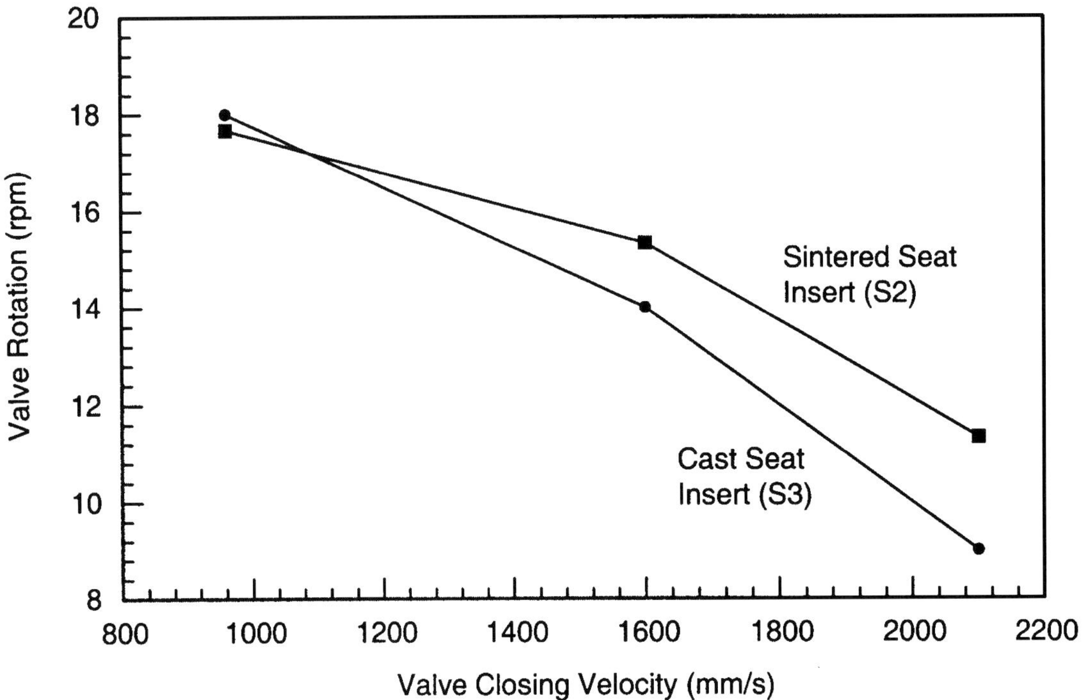

Fig. 6.23 Valve rotation against valve closing velocity for a V1 valve run against an S2 sintered seat insert and a V1 valve run against an S3 cast seat insert on the motorized cylinder head. See Table 6.1 for details of materials

It is not possible to assess whether rotation influenced valve or seat insert wear in the motorized cylinder head as valve rotation could not be varied while keeping other test parameters constant.

6.2.2.9 Effect of temperature

It has been shown that the general trend is that wear decreases as the temperature is increased from 150 to 600 °C [8]. This was thought to be because of oxide formation at high temperatures preventing metal-to-metal contact, and thus reducing adhesive wear. For this reason tests run on the hydraulic loading apparatus were not designed to study the effect of temperature on valve wear. The majority of the tests were run at room temperature. Running at an increased temperature of 130 °C using a hot air supply (see Fig. 5.2) was found to have little effect on the wear rate of the valves or seat inserts (as shown in Fig. 6.24). The valves and seat inserts exhibited similar wear features to those run at lower temperatures. It was not possible to run the hydraulic loading apparatus at temperatures higher than 130 °C.

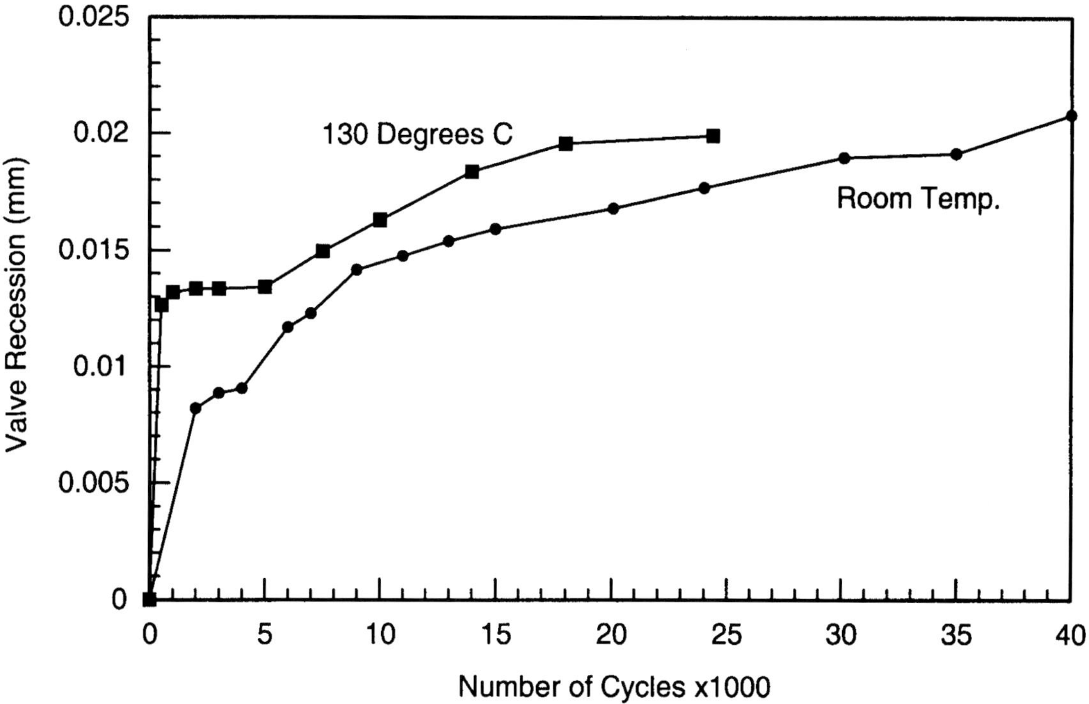

Fig. 6.24 Valve recession for V1 valves run at two different temperatures against S2 sintered seat inserts using impact and sliding on the hydraulic loading apparatus. See Table 6.1 for details of materials

At higher temperatures it is possible that the hardness of the valve or seat insert material may be reduced due to tempering, which could lead to increased valve recession or even fatigue failure of valves. Such temperatures are usually caused by a reduction in heat transfer from the valve head area as a result of deposit formation.

6.3 Seat insert materials

Having analysed the root causes of the valve recession mechanism and considered the implications for valvetrain design and seat insert and valve materials, it was decided to investigate the potential for reducing valve recession by improving existing seat insert materials for use in inlet valve applications, and to identify potential new seat insert materials. Reduction of valve recession can also be achieved by introducing design changes, but this is a more time-consuming process and would have implications for other aspects of engine performance.

The aim of this work was to test potential new seat insert materials and compare the results with those for existing materials. The wear mechanisms were investigated and the material performance was assessed. Testing was carried out using the bench test apparatus designed to simulate the loading environment and contact conditions to which a valve and seat insert are subjected (as described in Chapter 5).

6.3.1 Experimental details

6.3.1.1 Valve and seat insert materials

Selection of the potential new seat insert materials was based upon knowledge of the prevalent wear mechanisms and problems such as valve misalignment due to thermal distortions and valve seat loosening. The work described in Section 6.2 showed that the two main causes of valve recession are impact wear and frictional sliding and that the effect of both are increased by misalignment of the valve due to thermal distortions in the seat. Before selecting materials it was, therefore, necessary to identify the properties required to reduce the effect of these mechanisms.

One of the most important factors controlling impact wear has been shown to be fracture toughness [7]. The higher the fracture toughness of a material, the lower the impact wear. The effect of frictional sliding can be reduced by increasing the lubricity of the seat material, especially under dry running conditions. The problem of thermal distortion can be reduced by improving the conduction of heat through the seat area to the cooling channels in the cylinder head (a valve transfers approximately 75 per cent of the heat input to the top-of-head through its seat insert into the cylinder head [9]).

The most important properties required of a seat insert material in order to decrease the possibility of valve recession occurring were, therefore, considered to be:

- high fracture toughness;
- good lubricity under dry running conditions;
- high thermal conductivity.

The properties required of a seat insert material in order to reduce the likelihood of insert 'drop-out' occurring have been listed as [10]:

- high compressive yield strength;
- low modulus of elasticity;
- high thermal conductivity;
- normal thermal expansion.

Three materials were selected using these criteria. Each material does not fulfil all the criteria listed, but as each valve recession problem has its own set of operating parameters and design features it was thought best to test materials with a range of properties. Maraging steel was selected for its high resistance to impact. EN 1A, a free-cutting mild steel, was selected as it was thought the alloying elements incorporated to increase the machinability would help decrease the effect of frictional sliding while still giving good resistance to impact. Oilite, an oil-impregnated, porous, bronze material, was selected exclusively for its resistance to frictional sliding wear. Two types of cast iron were chosen to represent seats formed in-situ within the cylinder head: a grey cast iron and a ductile cast iron.

It was decided not to consider ceramic seat insert materials in this investigation as they are expensive to machine and the sponsor was more interested in low-cost options. It is possible that the wear properties of ceramic materials will be examined in the future. Results for the materials listed above were compared with those for two of the seat insert materials used in hydraulic loading apparatus work described in Section 6.2. The first was a sintered material (S2) consisting of a martensitic tool steel matrix with evenly distributed intra granular spheroidal alloy carbides, and the second a cast material (S3) consisting of a tempered martensitic tool steel matrix with a network of carbides evenly distributed.

All selected seat materials and their properties (where available) are shown in Table 6.6. Other selection criteria, such as cost and machinability, will be discussed further on.

Valves made from a martensitic, low-alloy steel (V1) were selected for use in the tests.

6.3.1.2 Specimen details

The geometries of the valves and seat inserts used are shown in Fig. 5.4. The materials not available as seat inserts and the cylinder head materials were made up into specimens as shown in Fig. 5.5. These could be used in both test rigs.

Table 6.6 Properties of the seat materials

Material	Fracture toughness ($MNm^{3/2}$)	Hardness (Hv)	Tensile strength (MN/m^2)	Thermal conductivity (W/m°C)	Coeff. of thermal expansion (μm/m°C)	Modulus of elasticity (GN/m^2)
Cast seat insert (S3)		490	250–400	40	10.3–12.6	120
Sintered seat insert (S2)		490				
Grey cast iron (250)	12	193	250	46	10.8 (20–400 °C)	110
Ductile cast iron (600)	20	319	600	32.5 (300 °C)	12.5 (20–400 °C)	174
Free-cutting mild steel (EN 1A)		205	360	43 (400 °C)	13.95 (20–400 °C)	185
Maraging steel (250)	120	358	1800	20.9 (100 °C)	10.1 (24–284 °C)	186
Oilite		89.4	96.5		17.64 (20–30 °C)	

6.3.1.3 Test methodologies

The impact and sliding test methodology (see Section 5.5.1.2) was utilized on the hydraulic loading apparatus in order to investigate the effect of the impact of the valve on the seat insert as the valve closes, in combination with the combustion loading. Test parameters used for selected tests are shown in Table 6.7.

Table 6.7 Seat insert material impact and sliding test parameters

Seat insert material	Valve temp. (°C)	Freq. (Hz)	Valve lift (mm)	Valve closing velocity (mm/s)	Load (kN)	Displacement waveform	Misalignment (mm)	No. of cycles	Lubn. (Y/N)
Grey cast iron (250)	R.T.	10	0.6	18	13	Sinusoidal	0	100 006	N
Ductile cast iron (600)	R.T.	10	0.6	18	13	Sinusoidal	0	100 013	N
Free cutting mild steel (EN 1A)	R.T.	10	0.6	18	13	Sinusoidal	0	100 021	N
Maraging steel (250)	R.T.	10	0.6	18	13	Sinusoidal	0	100 074	N
Oilite	R.T.	10	0.6	18	13	Sinusoidal	0	100 212	N
Grey cast iron (250)	R.T.	10	0.6	18	18.5	Sinusoidal	0	100 008	N
Ductile cast iron (600)	R.T.	10	0.6	18	18.5	Sinusoidal	0	100 107	N

In order to investigate the effect of impact of the valve on the seat insert materials as the valve closes, tests were run on the motorized cylinder head using seats manufactured from grey cast iron, ductile cast iron, free-cutting mild steel, maraging steel, and oilite. The same clearance was used for each valve, giving a constant closing velocity for each material. Details of the clearance used and the closing velocity, energy, and force are shown in Table 6.8. The tests were run for 100 000 cycles. Valve rotation was measured for each valve during the tests.

Table 6.8 Motorized cylinder head test parameters

Valve clearance (mm)	0.515
Closing velocity (mm/s)	1600
Closing energy (J)	0.234

6.3.2 Results

Figure 6.25 shows valve recession (calculated from wear scar data) from tests run on the hydraulic loading apparatus using *impact and sliding* for the seat materials listed in Table 6.6. The materials selected mainly for their resistance to frictional sliding exhibited the largest amount of recession. Of these, free-cutting mild steel provided the best performance, but this material was expected to have a greater resistance to impact than grey cast iron and oilite. It is clear that resistance to impact wear is a key material property in reducing the likelihood of valve recession.

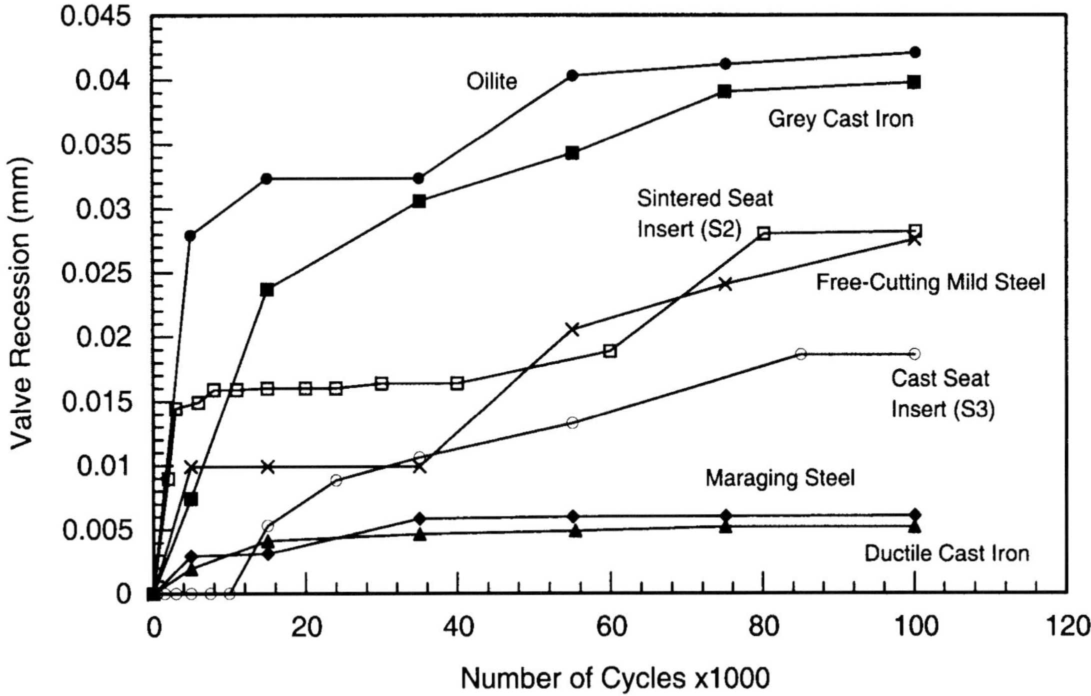

Fig. 6.25 Valve recession for V1 valves run against various seat materials using impact and sliding on the hydraulic loading apparatus

The ductile cast iron recessed less than might have been expected. Given its low fracture toughness compared to maraging steel, a higher recession relative to this material would have been anticipated. The low wear may be explained by its higher hardness combined with the presence of graphite, which can act as a solid lubricant, providing improved resistance to sliding wear compared with maraging steel.

It was not, however, unexpected that the ductile cast iron recessed less than the grey cast iron and the free-cutting mild steel. Additives, introduced into the molten iron just before casting of ductile cast iron, cause the graphite to grow as spheres rather than flakes, as in grey cast iron. Ductile cast iron is consequently stronger and more ductile than grey cast iron, giving it an increased resistance to impact.

On examination of the wear scars from all the tests it was found that the material combinations exhibiting the greatest recession gave the least severe valve wear and the most severe seat wear.

Wear features such as radial indentations were again observed on seat seating faces. On examination of the oilite seat before testing it was found that the seating face was covered in cracks and in places chunks of the material had been torn away on machining. However, after testing it was found that the cracks had disappeared. The only explanation was the occurrence of material flow at the seating face. Observation of the valve wear scar revealed the presence of scratches in the radial direction across the entire width, not observed with any other material combination.

Figure 6.26 illustrates the increase in recession recorded when increasing the combustion loading while maintaining the valve closing velocity, using a ductile cast iron seat. Again radial indentations on the seat seating face increased in number at the higher load.

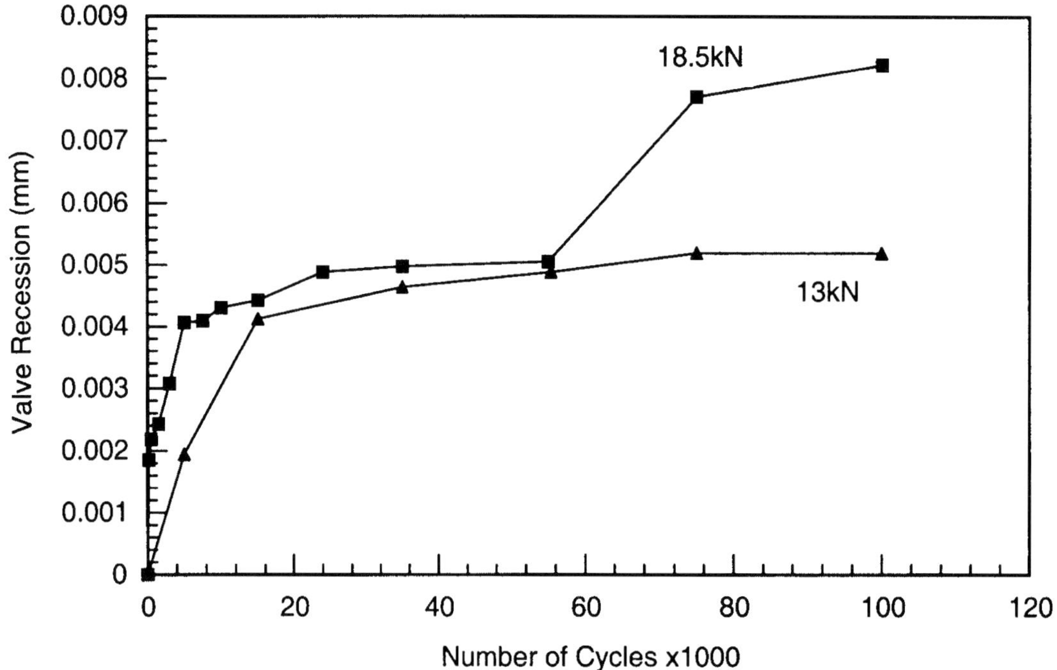

Fig. 6.26 Valve recession for V1 valves run at two different combustion loads against ductile cast iron seats using impact and sliding on the hydraulic loading apparatus

Figure 6.27 shows the valve recession (calculated from wear scar data) from tests run on the motorized cylinder head. Maraging steel and free-cutting mild steel have not been included as they exhibited exceptionally high recession in just the first 50 000 cycles (0.295 mm and 0.23 mm, respectively). They were both located in the seat position furthest from the driving belt and pulley system (see Fig. 5.12). It has been speculated that the unexpected level of impact wear was caused by torsional vibration of the camshaft, which only affected this seat position. In normal operation, in an engine, the camshaft would have a balancing flywheel at the opposite end to the driven end to eliminate such vibrations, but this was not present on this rig. The follower at this seat position eventually disintegrated and damage was also observed on the cam. This had not been a problem on previous tests as this valve position had not been utilized. Unfortunately, time constraints meant the tests could not be repeated to obtain more data for the two materials or to further study the effect. This could give cause, however, to believe that valve position influences the magnitude of wear. Instances where one valve has recessed far more than the others in the same cylinder head have been observed during engine testing (as described in Section 1.2). The cause, however, was not identified. Further work would be required to establish whether valve position affects recession and, if so, the mechanism which leads to its occurrence.

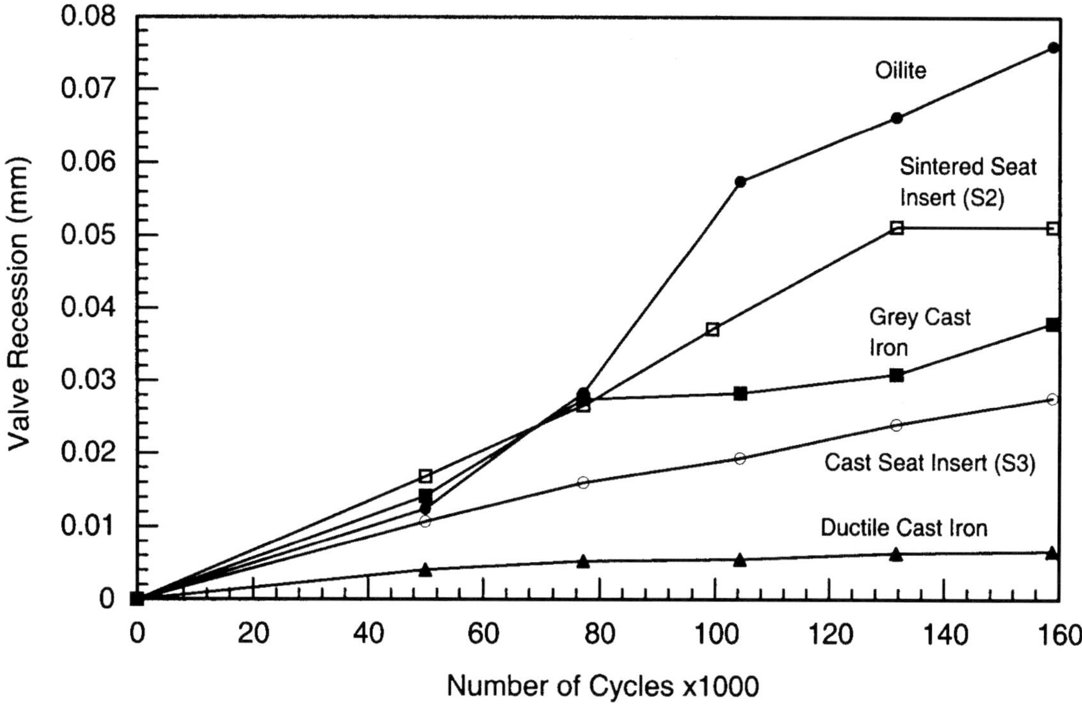

Fig. 6.27 Valve recession for V1 valves run against various seat materials with a valve closing velocity of 1600 mm/s on the motorized cylinder head

It can be seen that, relatively, the wear follows the same pattern on the motorized cylinder head as it did on the hydraulic loading apparatus, with one exception. On the hydraulic loading apparatus, more recession was recorded for the grey cast iron seat than the sintered seat insert, whereas on the motorized cylinder head the sintered seat insert exhibited greater wear. Clearly the sintered material is less resistant to impact wear than grey cast iron, but more resistant to sliding wear. A similar relative position for the ductile cast iron verifies that as expected it has high resistance to impact. This explains the good performance shown by this material in the hydraulic loading apparatus.

An accurate relationship between valve recession and seat hardness or seat fracture toughness could not be identified from the range of materials tested. With the little fracture toughness data for the materials available and ignoring the low recession exhibited when using a ductile cast iron seat, it could be said that wear decreases as fracture toughness increases. Ignoring the S2 sintered seat insert material, recession could be said to be inversely proportional to seat hardness.

As well as looking at resistance to impact and sliding in the materials selection process, machinability and cost of materials must also be considered. Some of the properties which give low tool wear and hence good machinability, listed by Mills and Redford [11], are:

1. low yield strength and low work hardening rate;
2. high thermal conductivity;
3. low chemical reactivity with tool or atmosphere;
4. low fracture toughness.

Clearly several of these properties are in direct contradiction to those required of the valve/seat materials in order to resist impact and sliding wear. It was noted that while, generally, for a group of similar materials machinability improves as the fracture toughness of the workpiece reduces, there are exceptions. Spheroidal (ductile) cast irons having higher fracture toughness than similar flake graphite cast irons actually give lower cutting-tool wear rates.

The efficiency with which materials are machined can be measured effectively by assessing the power required to machine a unit volume of material in unit time (specific power consumption). Data in the *Metals Handbook of Machining* [12] gives power required for machining for a range of engineering materials. Table 6.9 gives approximate powers for the seat materials tested. Approximate costs per kilogram for the materials [13] are also shown in Table 6.9.

Considering all the factors discussed (wear resistance, machinability, and cost) two materials stand out as potential seat/seat insert materials. These are maraging steel and ductile cast iron. Both gave far higher wear resistance than the two seat insert materials tested. While more power is consumed machining the ductile cast iron it is relatively cheap compared to the other materials. Maraging steel is, however, both expensive and difficult to machine. Ductile cast iron has the added advantage in that it can be alloyed with small amounts of nickel, molybdenum, or copper to improve its strength and hardenability. Larger amounts of silicon, chromium, nickel, or copper can be added for improved resistance to corrosion or for high temperature applications. This may mean it could also be suitable for exhaust valve applications, making it a viable option for use in manufacturing the entire cylinder head.

Table 6.9 Approximate power consumption and costs for the seat materials

Seat material	Power required for turning (Watts per mm^3 per min $\times 10^4$) [1]	Cost per kg (£) [2]
Oilite	135	1.1–1.4
Maraging steel	420	1.2–1.8
Free-cutting mild steel	210	0.25–0.35
Grey cast iron	180	0.2–0.35
Ductile cast iron	540	0.2–0.35

[1] ASM [12]

[2] Granta Design Ltd [13]

6.4 Conclusions

Stating the root cause of valve recession is difficult. Each valve recession problem will have its own unique set of operating parameters, design features, and material combinations. The investigation, however, has clearly shown that:

1. The bench test apparatus provides a valid simulation of the wear of both inlet valves and seat inserts used in automotive diesel engines.
2. The valve and seat insert wear problem involves two distinct mechanisms: impact of the valve on the seat insert on valve closure and sliding of the valve on the seat insert under the action of the combustion pressure.
3. Wear increases with valve closing velocity, combustion load, and misalignment of the valve relative to the seat insert. When using S3 cast seat inserts, valve recession was proportional to the square of the closing velocity (see Fig. 5.21).
4. Lubrication of the valve/seat interface reduced valve recession, on the material combination tested, by a factor of 3.5.
5. Resistance to impact is a key material property in reducing the likelihood of valve recession.
6. Considering all the factors discussed (resistance to impact and sliding wear, machinability and cost), two materials stand out as potential seat/seat insert materials. These are maraging steel and ductile cast iron. Both gave far higher wear resistance than the two seat insert materials tested. While more power is consumed machining the ductile cast iron it is relatively cheap compared to the other materials. Maraging steel is, however, both expensive and difficult to machine.
7. Data is available for the development of a semi-empirical model for predicting valve recession.

6.5 References

1. **Van Dissel, R., Barber, G.C., Larson, J.M.,** and **Narasimhan, S.L.** (1989) Engine valve seat and insert wear, SAE Paper 892146.
2. **Zum-Gahr, K.H.** (1987) *Microstructure and wear of materials*, Tribology Series No. 10, Elsevier, Amsterdam.
3. **Fricke, R.W.** and **Allen, C.** (1993) Repetitive impact-wear of steels, *Wear*, **163**, 837-847.
4. **Hutchings, I.M., Winter, R.E.,** and **Field, J.E.** (1976) Solid particle erosion of metals: the removal of surface material by spherical objects, *Proc. R. Soc. Lond. Ser. A.*, **348**, 379-392.
5. **Lewis, R.** (2000) *Wear of diesel engine inlet valves and seats*, PhD Thesis, University of Sheffield, UK.
6. **Rabinowicz, E.** (1995) *Friction and wear of materials*, Second Edition, John Wiley and Sons, New York.

7. **Kawachi, R., Tujii, H., Kawamoto, M.,** and **Okabayashi, K.** (1983) On the impact wear of carbon steel and cast iron, *J. Japan Inst. Metals*, **47**, 225–230.

8. **Wang, Y.S., Narasimhan, S., Larson, J.M., Larson, J.E.,** and **Barber, G.C.** (1996) The effect of operating conditions on heavy duty engine valve seat wear, *Wear*, **201**, 15–25.

9. **Giles, W.** (1971) Valve problems with lead free gasoline, SAE Paper 710368.

10. **Lane, M.S.** and **Smith, P.** (1982) Developments in sintered valve seat inserts, SAE Paper 820233.

11. **Mills, B.** and **Redford, A.H.** (1983) *Machinability of engineering materials*, Applied Science Publishers, London.

12. **ASM** (1967) *Metals handbook: Vol. 3 – Machining*, ASM International, Ohio.

13. **Granta Design Ltd** (1994) Cambridge materials selector software, Version 2.02.

Chapter 7

Design Tools for Prediction of Valve Recession and Solving Valve Failure Problems

7.1 Introduction

An important aspect of research is the industrial implementation of the results. This chapter looks at the provision of tools that will enable the results of the review of literature, analysis of failed specimens, and bench test work to be applied in industry to assess the potential for valve recession and solve problems that occur more quickly.

The development of a semi-empirical model for predicting valve recession is described. Flow charts outline steps to be used to reduce the likelihood of recession occurring during the design process, as well as offering solutions to problems that do occur.

The model developed for predicting valve recession was based on the mechanisms of valve and seat wear determined during investigations carried out using purpose built valve wear test apparatus (as detailed in Chapter 6). Existing models for the mechanisms of wear observed were combined to produce the final model. Constants required for the model were taken from curves fitted to experimental data. An iterative software program called RECESS[1] was developed to run the model. The model was then validated against both engine tests and tests run on the hydraulic loading apparatus.

The flow charts were developed from data collected from the review of literature, analysis of failed specimens, and modelling.

7.2 Valve recession model

7.2.1 Review of extant valve wear models

Existing models developed for assessing the likelihood of the occurrence of valve wear have been simplistic. They focussed on the frictional sliding between the valve and seat

[1] RECESS is a commercially available software package and database. Contact the authors for further details.

under the action of the combustion pressure and took no account of the impact of the valve on the seat on valve closure.

The valve wear factor *Fw* derived by Pope [1] for medium-speed diesel engines, equation (7.1) varies directly with the coefficient of friction. Other variables are dependent on engine design. It was found that a wear factor of 200 gave a satisfactory service life and wear factors greater than 250 gave a poor service life. The length of a 'satisfactory life', however, was not defined.

$$Fw = \frac{\mu \, P^2 \; N \; D^5 \; \tan\theta}{E \; B \; C \; t \; t_m^2} \tag{7.1}$$

where μ is the coefficient of friction between valve and seat, P is the maximum cylinder pressure (psi), N is the engine speed (r/min), D is the valve disc diameter (in), θ is the seat angle (degrees), E is the Young's modulus for the valve material (psi), B is the Rockwell Hardness number, C is the height of seat (in), t is the distance from the valve disc face to the seat top (in), and t_m is the height of the valve disc cone (in). Details of the valve dimensions required are shown in Fig. 7.1.

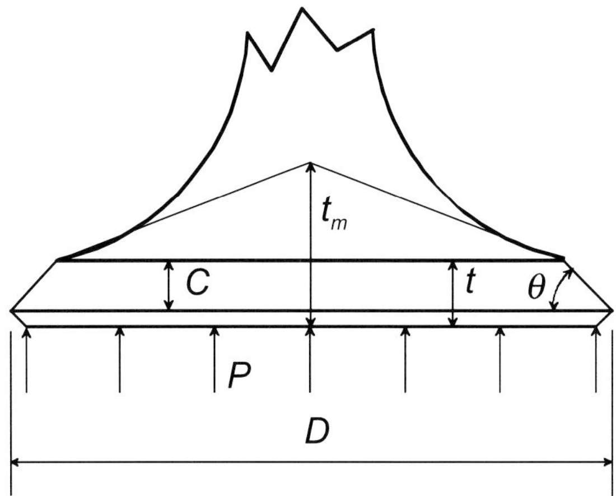

Fig. 7.1 Valve details [1]

For a N/A, 1.8 litre, IDI, diesel engine inlet valve and seat (valve material V1 and seat insert material S3) $P = 1885.5$ psi; $\theta = 45$ degrees; $C = 0.087$ in; $N = 4800$ r/min; $E = 29\,007\,367.9$ psi; $t = 0.122$ in; $D = 1.417$ in; $B = 79.5$ RA; $t_m = 0.481$ in; $\mu = 0.4$ (steel on steel, dry) or 0.05 (steel on steel, greasy).

Therefore, for a 'dry' valve/seat insert contact $Fw = 3154$ and for a 'greasy' valve/seat insert contact $Fw = 394$, both of which lie in the range likely to give an 'unsatisfactory' life. Engine test data (shown in Fig. 4.4), however, clearly indicate that wear falls within acceptable boundaries for this material combination.
Giles [2] used the following expression, based on the equation developed by Archard [3] for predicting volume of material lost through adhesive wear V, to investigate ways in which valve recession could be reduced

$$V = \frac{KWL}{CP} \tag{7.2}$$

where K is the wear coefficient, W is the load on sliding surfaces, L is the sliding distance, P is the penetration hardness of the softer of the two contacting surfaces, and C is a constant depending on units of measurement.

By admission, however, this was only intended to provide a generalized approach to minimizing wear rather than perform a quantitative analysis.

It was clear that a model was required that would provide a quantitative analysis of valve recession. In order to achieve this it needed to encompass all wear mechanisms occurring at the valve/seat interface. To allow a range of inlet valve and seats to be considered, valve parameters, design parameters, and material properties also needed to be included.

7.2.2 Development of the model

Development of the model followed the stages detailed in the flow chart shown in Fig. 7.2. It was decided to consider the two fundamental wear mechanisms identified as causing valve recession separately (frictional sliding at the valve/seat insert interface under the action of the combustion pressure, and impact of the valve on the seat insert on valve closure) as they occur as two definable events in the valve operating cycle. Approaches for modelling the wear mechanisms identified for each were appraised and parameters were then derived either from test results or directly from the valve and seat design and engine operating conditions. An allowance for effects such as misalignment and variation in lubrication was also built in at this stage. The two parts were then combined to form the final valve recession model.

7.2.2.1 Frictional sliding

Abrasive, adhesive, and fretting wear were observed to have occurred as a result of the frictional sliding between the valve and seat under the action of the combustion pressure. It was, therefore, decided to use the equation developed by Archard [3], for deriving wear volume V in sliding situations, to model the wear caused by the frictional sliding

$$V = \frac{kP_c x}{h} \tag{7.3}$$

where k is the wear coefficient, P_c is the contact force (N), x is the sliding distance (m) and h is the penetration hardness of the softer of the two contacting surfaces (N/m²).

Equation (7.3) was originally developed for modelling adhesive wear, but it has also been used successfully to model both abrasive wear [4] and fretting wear [5]. Calculation of the parameters to be used in equation (7.3) is outlined below.

Load

The peak load normal to the direction of sliding at the valve/seat insert interface P_c was calculated using the peak combustion pressure p_p and the valve head geometry, as shown in Figure 7.3

$$P_c = \frac{p_p \pi R_v^2}{(1+\mu)\sin\theta_v} \tag{7.4}$$

where θ_v is the valve seating face angle (degrees) and μ is the coefficient of friction at the valve/seat interface.

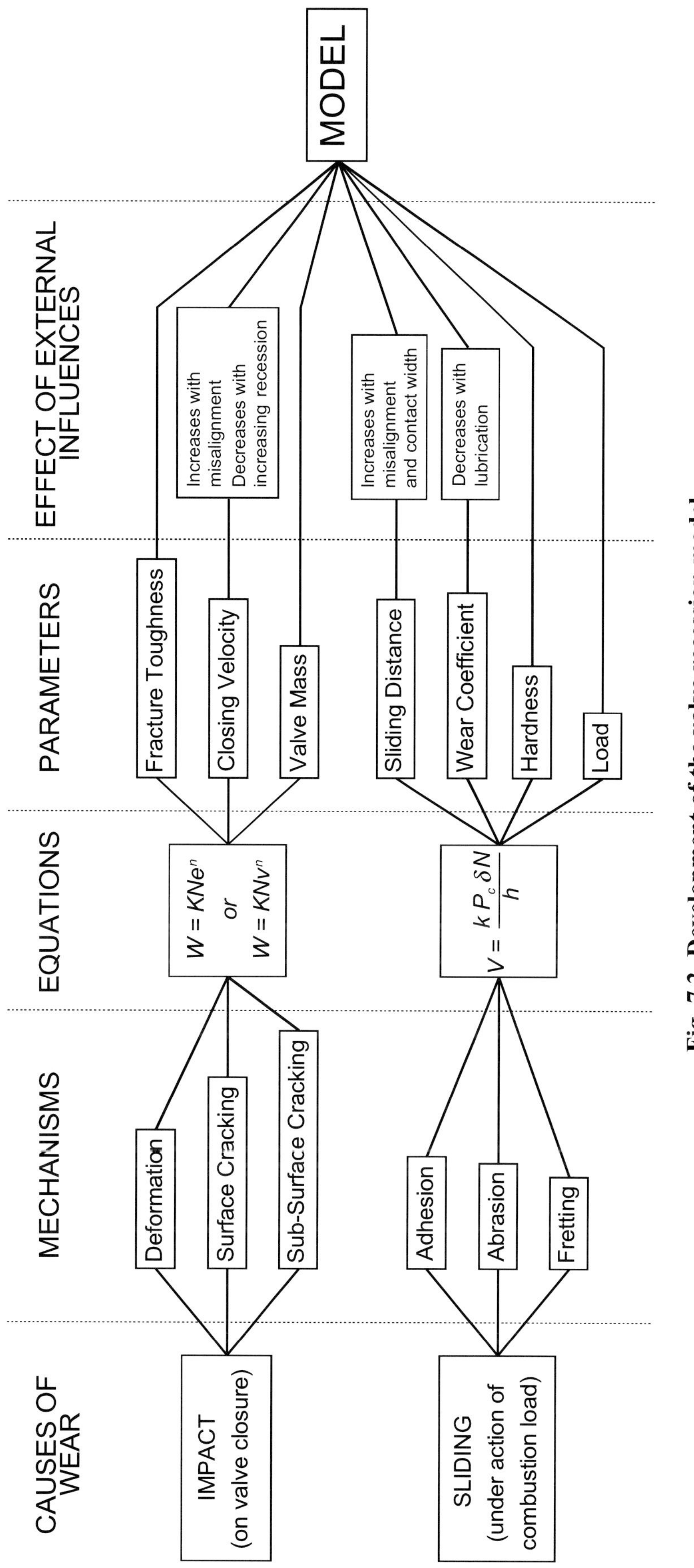

Fig. 7.2 Development of the valve recession model

The load on the valve seat is initially zero then rises to P_c and falls back down to zero. For the purposes of calculating the sliding wear volume an average load $\overline{P}$ was assumed equal to half P_c. In the absence of other data, μ was estimated to be 0.1 for the valve/seat interface, which is a typical value for boundary lubricated steel surfaces.

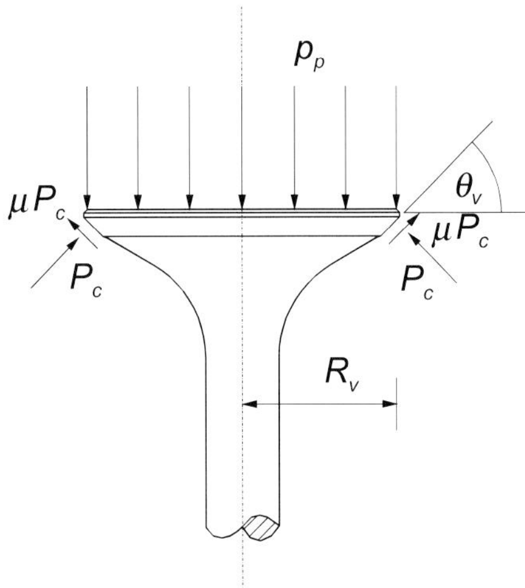

Fig. 7.3 Valve geometry and loading

Sliding distance

Slip at the interface can be found either by measurement of scratches or by using finite element analysis. Data generated by Mathis *et al.* [6] using the latter were utilized in this model. In the absence of other data it was assumed that slip at the interface δ is proportional to combustion load P_c. The total sliding distance is calculated by multiplying the slip δ by the number of loading cycles N.

Wear coefficient

Rabinowicz [7] carried out a series of studies to systematically generate wear coefficients for sliding metals. These were derived from wear volumes using equation (7.3). Different states of lubrication and material 'compatibility' ('the degree of intrinsic attraction of the atoms of the contacting metals for each other, as demonstrated by the solid solubility or liquid miscibility' [7]) were investigated. The published data were used to construct a chart of wear coefficients, as shown in Fig. 7.4. Adhesive wear data are shown as well as data applicable to abrasive, fretting, and corrosive wear. The lubrication states and material compatibilities for valve/seat insert material pairings used in both engine tests and rig tests are shown in Table 7.1. The wear coefficients, also listed, were taken from Fig. 7.4. It should be noted that the wear coefficients given in Fig. 7.4 were derived using the form of Archard's equation in which wear is proportional to sliding distance multiplied by load divided by three times the hardness. In this work the three was taken out and the wear coefficients adjusted accordingly.

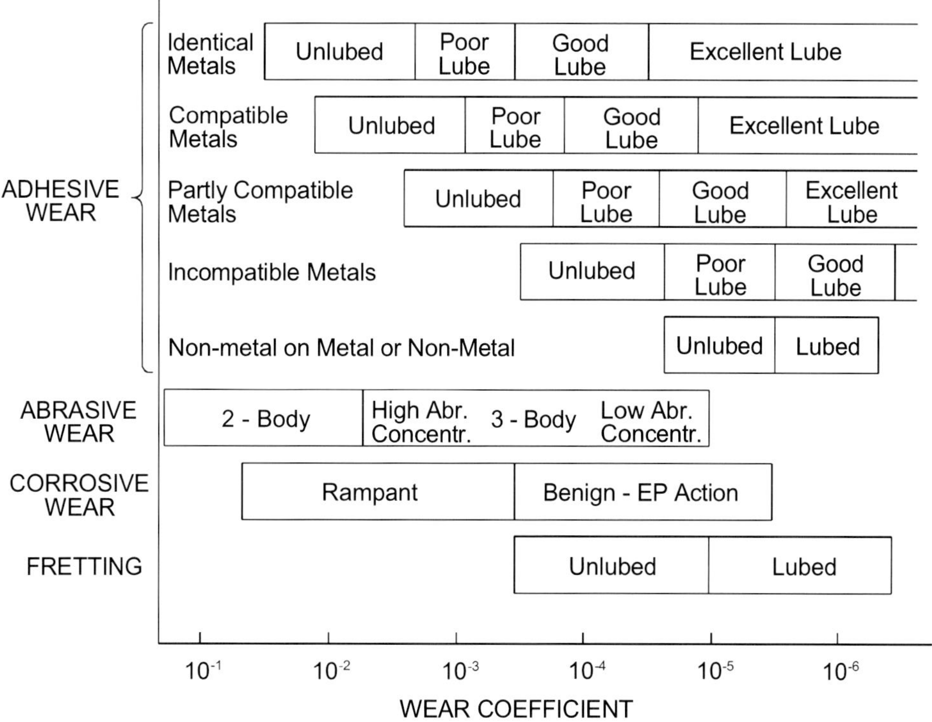

Fig. 7.4 Wear coefficients to be anticipated in various sliding situations [7]

Hardness

The penetration hardness of the softer of the two materials (valve and seat insert) was used. 'Hardness' of the valve and seat materials used during engine and rig tests are shown in Tables 6.2 and 6.6.

Table 7.1 Lubrication states, material compatibilities, and wear coefficients for valve/seat insert material pairings used in both engine tests and rig tests

Test apparatus	Valve material	Seat material	Lubrication	Material compatibility	Wear coefficient
Engine	V1	S1	Poor	Compatible	5×10^{-5}
Engine	V1	S3	Poor	Compatible	5×10^{-5}
Hyd. loading app.	V1	S2	Unlubed	Compatible	1×10^{-3}
Hyd. loading app.	V1	S3	Unlubed	Compatible	1×10^{-3}
Hyd. loading app.	V1	S3	Good	Compatible	1×10^{-6}
Hyd. loading app.	V1	Grey cast iron	Unlubed	Compatible	1×10^{-3}
Hyd. loading app.	V1	Ductile cast iron	Unlubed	Compatible	1×10^{-3}
Hyd. loading app.	V1	Free-cutting mild steel	Unlubed	Compatible	1×10^{-3}
Hyd. loading app.	V1	Oilite	Poor	Intermediate	5×10^{-5}
Hyd. loading app.	V1	Maraging steel	Unlubed	Compatible	1×10^{-3}

7.2.2.2 Impact

The observed deformation on the valve seating faces and surface cracking on the seat inserts are characteristic of processes leading to wear by single or multiple impact of particles [8] (see Section 6.2.2.1).

Fricke and Allen [9] used a relationship of the same form as that used in erosion studies to model impact wear of poppet valves operating in hydropowered stoping mining equipment

$$W = KNe^n \tag{7.5}$$

where W is the wear mass (Kg), e is the impact energy per cycle (J) and K and n are empirically determined wear constants and

$$e = \frac{1}{2}mv^2$$

where m is the mass of the valve added to the mass of the follower and half the valve spring mass (kg), and v is the valve velocity at impact (m/s).

Wellinger and Breckel [10], in their repetitive impact studies, also found that wear loss could be described by

$$W = KNv^n \tag{7.6}$$

Fricke and Allen [9] justified the use of such a relationship for impact wear of valves, citing work by Hutchings *et al.* [11] in which it was shown that erosion can be satisfactorily modelled by the impact of large particles. In their work they used hard steel balls up to 9.5 mm in diameter. It was thus assumed that a relationship exists between impacts on a macroscale (greater than 1 mm) and impacts on a microscale (less than 1 mm), such as those found typically in erosive wear. The similarity of the wear features observed during testing to those attributed to erosive wear further supports this approach (see Section 6.2.2.1) as well as the fact that wear was found to be approximately proportional to the square of the closing velocity (see Section 6.2.2.7). It was, therefore, decided that an equation of this form would be appropriate to model the wear caused by the impact of the valve on the seat insert on valve closure. Although the valve closing velocities achieved using the hydraulic loading apparatus were not representative of those found in an engine, equation (7.5) was thought to be more suitable as the energies on valve closure were quite close to those found in an engine (see Section 5.5.3.2). Calculation of the parameters used in equation (7.5) is outlined below.

Velocity

Valve closing velocity was determined from the initial valve clearance c_i using the velocity curves illustrated in Section 5.5.3.1. This was achieved by first fitting a line to the valve lift curve (fourth-order polynomial) to obtain lift as a function of time and then differentiating this equation to give velocity as a function of time. Velocity was then plotted against lift and a further line fit gave velocity as a function of lift.

As the valve recesses, the valve clearance at closure will decrease, thereby decreasing the valve closing velocity. This needs to be considered in the application of the model.

Wear constants K and n

Values of K and n for the seat materials used in these studies were derived using an iterative process to fit equation (7.5) to experimental data from tests run on the motorized cylinder head. Equation (7.5) was used to calculate wear volumes rather than wear mass to fit in with Archard's sliding wear equation, equation (7.3), when combining the two to create the final model. These were then used to calculate recession values using equations relating wear volume to recession [12]. At each data point the velocity was recalculated to take account of the change in clearance due to recession. An iterative process was used to fit the recession values to those derived for tests run on the motorized cylinder head. The results of this process for tests run with V1 valves and S3 cast seat inserts, V1 valves and S2 sintered seat inserts, and V1 valves and grey cast iron, ductile cast iron, and oilite seats are shown in Figs 7.5, 7.6, and 7.7 respectively.

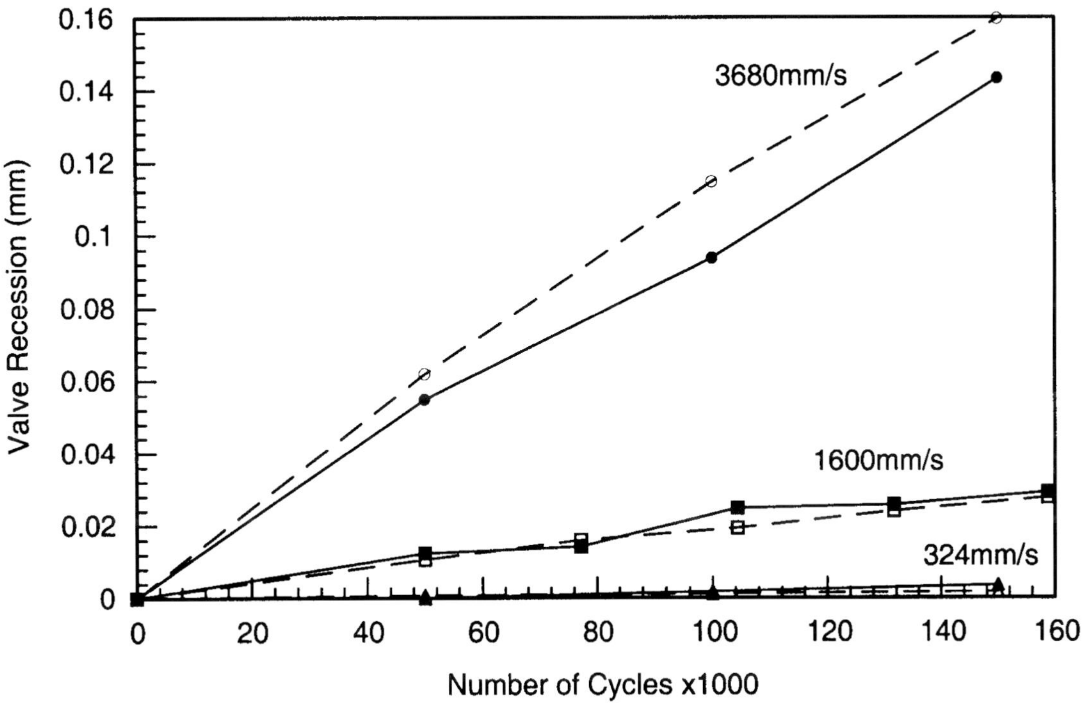

Fig. 7.5 Modelling of valve recession for V1 valves run with three different closing velocities against S3 cast seat inserts on the motorized cylinder head (solid line – experimental data; broken line – model prediction)

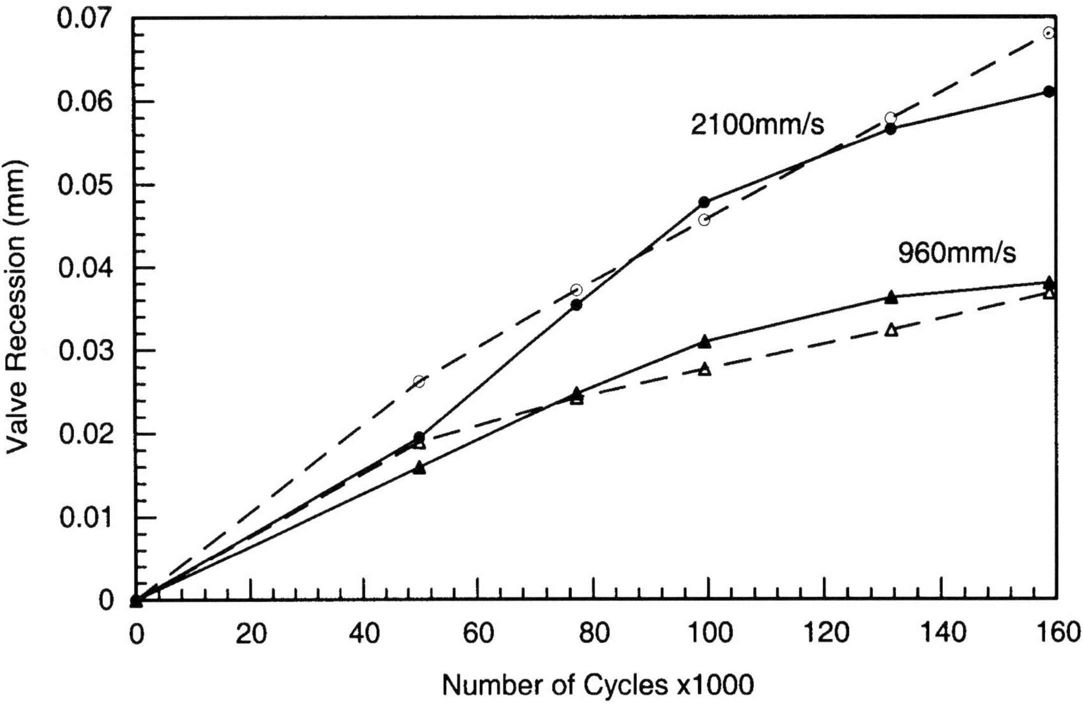

Fig. 7.6 Modelling of valve recession for V1 valves run with two different closing velocities against S2 sintered seat inserts on the motorized cylinder head (solid line – experimental data: broken line – model prediction)

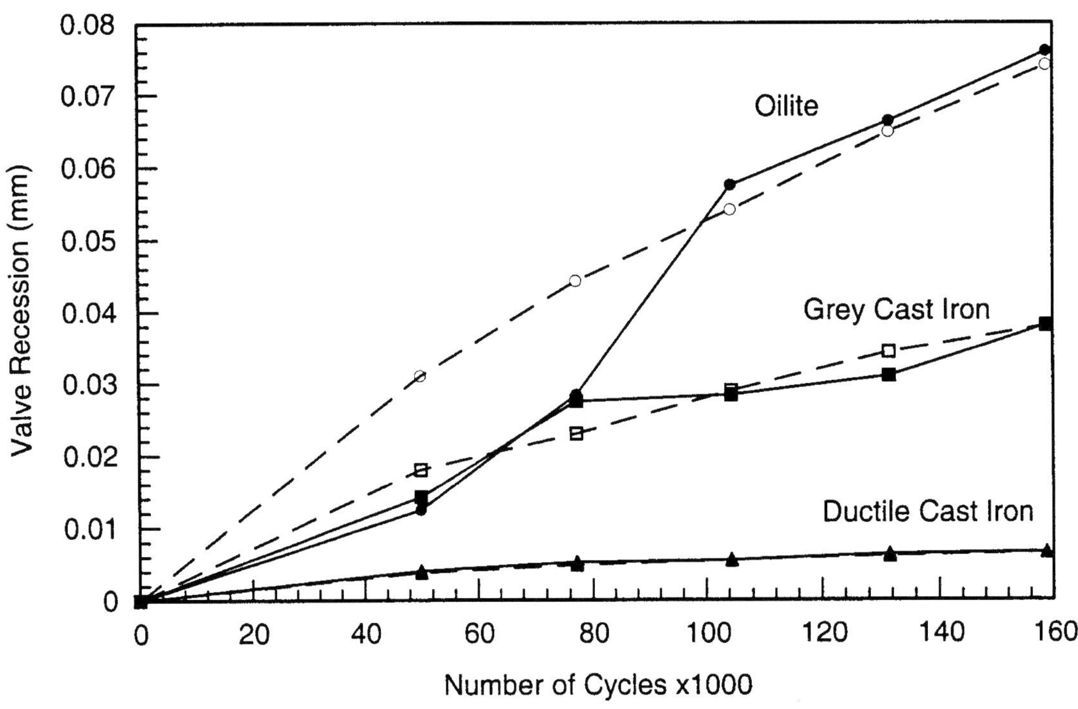

Fig. 7.7 Modelling of valve recession for V1 valves run against various seat materials with a closing velocity of 1600 mm/s on the motorized cylinder head (solid line – experimental data; broken line – model prediction)

As can be seen for seat insert materials S1 and S2, the values of K and n derived give good correlation over a range of velocities. This proves that the impact model is relevant.

The values of K and n used to produce the model predictions shown in Figs 7.5–7.7 are listed in Table 7.2.

Table 7.2 Values of impact wear constants *K* and *n* for valve/seat insert material pairings

Valve Material	Seat Material	K	n
V1	S3	5.3×10^{-14}	1
V1	S2	3.5×10^{-14}	0.3
V1	Grey cast iron	4.8×10^{-14}	0.77
V1	Ductile cast iron	5×10^{-14}	1.2
V1	Free-cutting mild steel	*	*
V1	Oilite	5×10^{-14}	0.2
V1	Maraging steel	*	*

* Test incomplete (see Section 6.3.2)

It was hoped that a relationship between the impact wear factor n and seat fracture toughness would emerge. However, the low recession exhibited by the ductile cast iron seats provided an anomaly. In general it could be said that n reduces with increasing seat hardness and seat fracture toughness. Without being more exact it would be impossible to build in such a relationship to the model. This means that testing on the motorized cylinder head is required in order to model the recession likely to occur with a particular material combination.

7.2.2.3 Final model

Putting together the equations for sliding and impact wear [(7.3) and (7.5)] gives the final wear model as

$$V = \frac{k\bar{P}N\delta}{h} + KNe^{n} \tag{7.7}$$

In order to incorporate the change in pressure at the interface and any other effects likely to lead to a reduction in the wear rate with time, a term consisting of the ratio of the initial valve/seat contact area A_i to the contact area after N cycles, A, to the power of a constant j was included. The term j was determined empirically using bench and engine test data.

$$V = \left(\frac{k\bar{P}N\delta}{h} + KNe^{n}\right)\left(\frac{A_i}{A}\right)^{j} \tag{7.8}$$

Equation (7.8) gives a wear volume which is then converted to a recession value using equations derived previously using the valve and seat geometries [12]. Equation (7.9) gives the geometrical relationship between r and V for the case where valve and seat angles are equal

$$r = \left(\sqrt{\frac{V}{\pi R_i \cos\theta_s \sin\theta_s} + w_i^2} - w_i \right) \sin\theta_s \tag{7.9}$$

where R_i is the initial seat insert radius, θ_s is the seat insert seating face angle, and w_i is the initial seat insert seating face width (as measured). R_i can be calculated using w_i and the radius and seating face width as specified for the seat insert (R_d and w_d).

Valve closing velocity is related to valve recession, which needs to be considered in the application of the model. Closing velocity decreases as recession increases.

7.2.3 Implementation of the model

A flow chart outlining the use of the model is shown in Fig. 7.8. The wear volume is determined incrementally. The initial valve closing velocity and contact area are used to calculate the volume of material removed over the first N cycles. This is then converted to recession, and new values for the clearance (and hence closing velocity) and contact area are determined. The calculation is repeated until the total number of iterated cycles equals the required run duration.

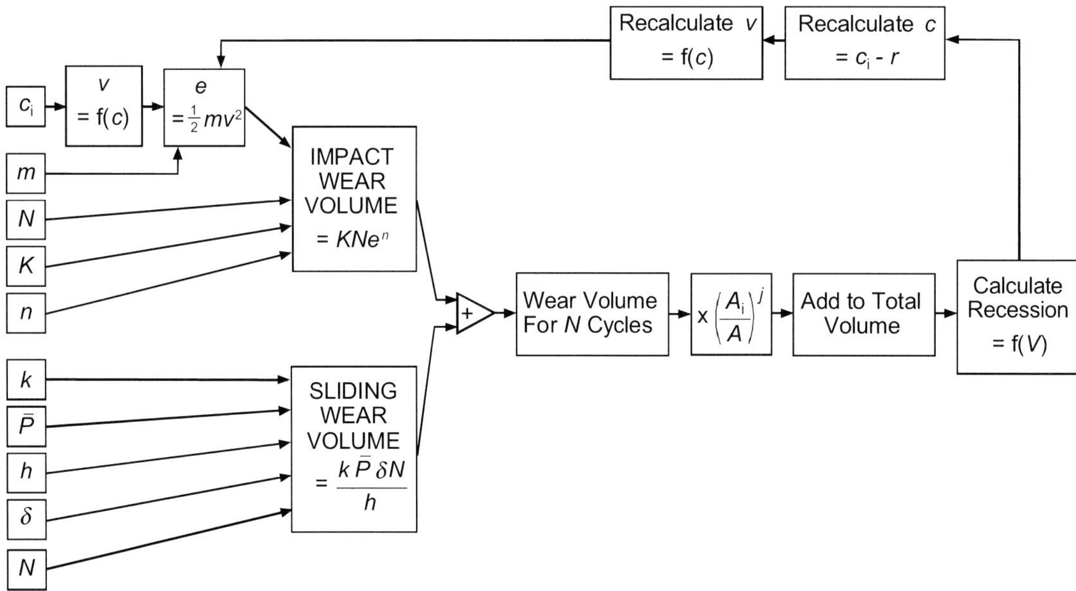

Fig. 7.8 Valve recession model application flow chart (Reprinted with permission from SAE paper 2001-01-1987 © 2001 Society of Automotive Engineers, Inc.)

In order to provide a tool for running the valve wear model, an iterative software program called RECESS was developed. Within the program the recession calculation is carried out in three stages. In the first and second, the sliding and impact wear volumes, V_s and V_i, are calculated for N cycles. In the final stage, V_s and V_i are added together to calculate the wear volume for the set of N cycles. This is then used to calculate the recession value r for the total number of cycles. The three stages are explained in more detail below.

Sliding wear calculation

At this stage, as shown in Fig. 7.9, the following are entered:

- valve and initial seat insert geometry;
- seat insert material properties;
- maximum combustion pressure, valve misalignment (if any), coefficient of friction at the valve/seat interface, wear coefficient for the material combination and the number of cycles.

The sliding wear volume is calculated for N cycles.

WEAR VOLUME DUE TO SLIDING		
Initial SI Seating Face Width (w_i) =	2.00E-03	m
Initial SI Radius (R_i) =	0.01685	m
Max. Combustion Pressure (p_p) =	2.00E+07	Pa
Valve Head Radius (R_v) =	1.80E-02	m
Seating Face Angle (θ_v)=	45	Degrees
Coeff. of Friction at Interface (μ) =	0.1	
Max. Contact Force at Interface (P_c) =	26172.61949	N
Avg. Contact Force at Interface ($P\ bar$) =	13086.30975	N
Valve Misalignment Relative to SI =	0	m
Slip at Interface (δ) =	8.06E-06	m
Number of Cycles (N) =	1440000	
Sliding Distance (x) =	11.6064	m
Wear Coefficient (k) =	1.00E-05	
Hardness of SI Material (h) =	4.90E+02	Hv (Kg/mm^2)
Wear Volume Due to Sliding (V_s) =	1.05324E-10	m^3

Fig. 7.9 Sliding wear calculation

Impact wear calculation

At this stage, as shown in Fig. 7.10, the following are entered.

- Valve clearance. Note that, for the 1.8 litre diesel inlet cam to be used in validating the model (see Section 7.2.4), valve clearances between 0 and 0.4 mm give a constant valve closing velocity and the first column should be used. For clearances above 0.4 mm the velocity varies with clearance. The clearance should then be entered in the second column where it is used to calculate a closing velocity using a curve fitted to clearance/closing velocity data for the cam.

- Mass of the valve, follower, and half the spring.
- Wear constants K and n (derived from experimental data).

Note that r at the top of the sheet should be left as zero for the first set of N cycles. After the first set of N cycles a value for r is calculated (at the *Recession* calculation stage, see below) which should then be entered for the subsequent set of N cycles in order to recalculate valve clearance and the closing velocity (not necessary for initial valve clearances below 0.4 mm).

The number of cycles N is taken directly from the *Sliding Wear* stage.

The impact wear volume is calculated for N cycles.

WEAR VOLUME DUE TO IMPACT			
	Clearance: 0 to 0.0004	*Clearance: 0.0004 +*	
Recession (r) =	0.00E+00	0	m
Valve Clearance (c) =	0.000235	0.0004	m
Valve Misalignment Relative to SI =	0	0	m
Total Clearance =	0.000235	0.0004	m
Velocity on Impact (v) =	0.288274	0.800492234	m/s
Mass (m) =	0.18295	0.18295	Kg
Energy on impact (e) =	0.007601746	0.058616065	J
Wear Constant K =	5.30E-14	5.30E-14	
Wear Constant n =	1	1	
Number of Cycles (N) =	1440000	1440000	
Wear Volume Due to Impact (V_i) =	5.80165E-10	4.47358E-09	m^3

Fig. 7.10 Impact wear calculation

Recession calculation

At this stage, as shown in Fig. 7.11, the value of j to be used is entered as well as the initial value of (A_i/A). The sliding wear volume V_s for the set of N cycles, and the impact wear volume V_i for the set of N cycles, are then added together to calculate the wear volume for the set of N cycles. This is multiplied by the $(A_i/A)^j$ term. This number should then be entered into the table on the sheet. The values in this table are added together to calculate the total wear volume V_t which is used to calculate the valve recession r for the total number of cycles, using the equation relating V_t to r for an equal valve and seat angle of 45 degrees [12].

The recession r calculated should then be re-entered at the *Impact Wear* stage in order to recalculate the valve clearance, c, and the valve closing velocity, v, for the next set of N cycles (not necessary for initial clearances below 0.4 mm), and the new value of (A_i/A) should be entered at the *Recession* stage.

RECESSION					
V_s (Wear Volume Due to Sliding) =	1.05324E-10	m^3	No of Cycles	Total Wear Volume	$(A_i/A)^j$
V_i (Wear Volume Due to Impact) =	5.80165E-10	m^3	N (1)	6.85E-10	1
$(A_i/A)^j$ =	0.762		N (2)	6.62E-10	0.966
Wear Volume for N Cycles =	6.8549E-10	m^3	N (3)	6.40E-10	0.934
Wear Volume for N Cycles * $(A_i/A)^j$	5.22343E-10		N (4)	6.20E-10	0.904
Total Wear Volume =	5.97E-09	m^3	N (5)	6.00E-10	0.876
Recession (r) =	3.93035E-05	m	N (6)	5.83E-10	0.851
Seat Insert Seating Face Width (w) =	0.002055584	m	N (7)	5.66E-10	0.826
SI Radius (R) =	0.016889304	m	N (8)	5.50E-10	0.803
Seating Face Area (A) =	0.000218136	m^2	N (9)	5.36E-10	0.782
Initial Seating Face Area (A_i) =	0.000211743	m^2	N (10)	5.22E-10	0.762
A_i/A =	0.9706955				
j =	10				
$(A_i/A)^j$ =	0.742728627				

Fig. 7.11 Recession calculation

7.2.4 Model validation

In order to validate the model valve, recession predictions were calculated for engine tests and tests run on the hydraulic loading apparatus.

7.2.4.1 Engine tests

When calculating recession predictions for engine test results, the model was used as described above. Equal valve and seat angles of 45 degrees and the same initial seating face widths were assumed. It was also assumed that the initial clearances were set within the constant velocity region (0–0.4 mm, see Fig. 5.14). Initial conditions used for the test simulated are shown in Table 7.3.

Table 7.3 Initial values used in calculating model predictions

Valve material	Seat material	Test type	w (mm)	k	P (N)	h (H^v)	v (mm/s)	K	n
V1	S1	Engine	2	5×10^{-5}	8053	490	288	3.5×10^{-14}	0.3
V1	S3	Engine	2	5×10^{-5}	8053	490	288	5.3×10^{-14}	1
V1	S3	Hyd. app.	0.719	1×10^{-3}	8053	490	18	5.3×10^{-14}	1
V1	S2	Hyd. app.	0.760	1×10^{-3}	8053	490	18	3.5×10^{-14}	0.3
V1	Grey C.I.	Hyd. app.	0.653	1×10^{v3}	8053	193	18	4.8×10^{-14}	0.77
V1	Ductile C.I.	Hyd. app.	0.796	1×10^{-3}	8053	319	18	5×10^{-14}	1.2
V1	Oilite	Hyd. app.	0.504	5×10^{-5}	8053	89.4	18	5×10^{-14}	0.2
V1	S3 (lubed)	Hyd. app.	0.726	1×10^{-6}	8053	490	18	5.3×10^{-14}	1

Figure 7.12 shows the model prediction for an engine test run using S3 cast and S1 sintered seat inserts. As can be seen, the model produces a good prediction of valve recession. In order to determine which parameters were having the largest influence on the wear, the percentage split of the total wear between impact and sliding for the engine test predictions was calculated (see Fig. 7.13). Clearly the contribution of impact wear to the total was higher than that of sliding. The parameters having the largest influence on the wear are, therefore, valve closing velocity, valve mass, and the impact wear constants K and n, which are related to the resistance to impact wear of a seat material.

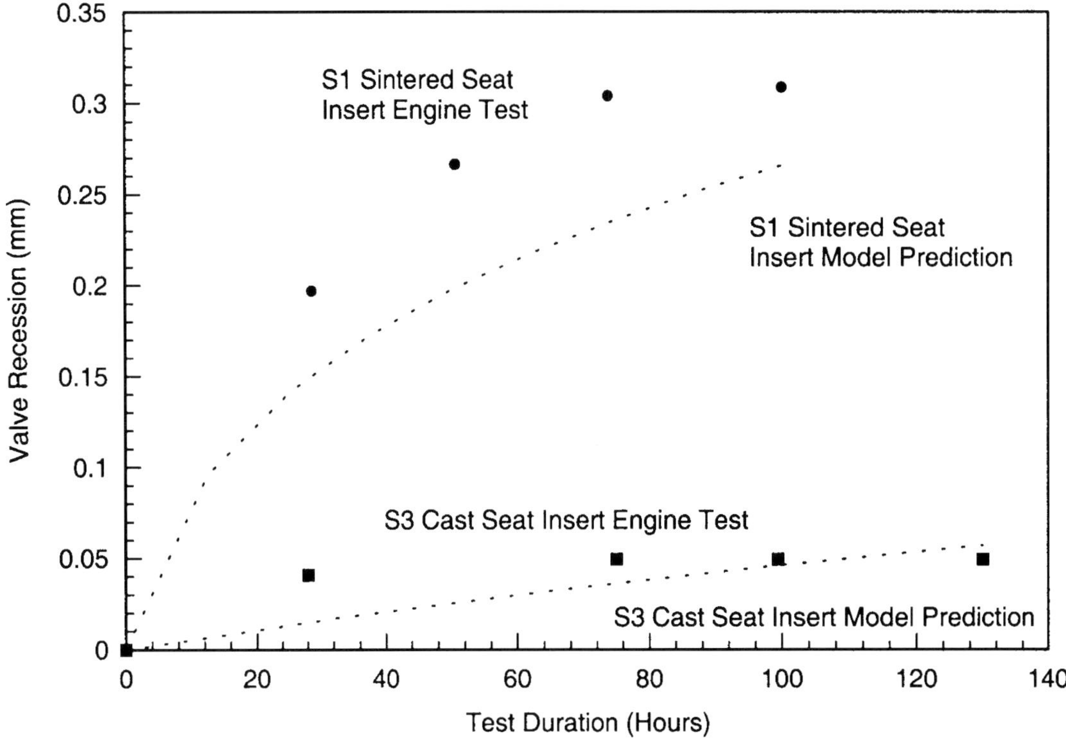

Fig. 7.12 Model prediction (broken line) versus engine test data. (Reprinted with permission from SAE paper 2001-01-1987 © 2001 Society of Automotive Engineers, Inc.)

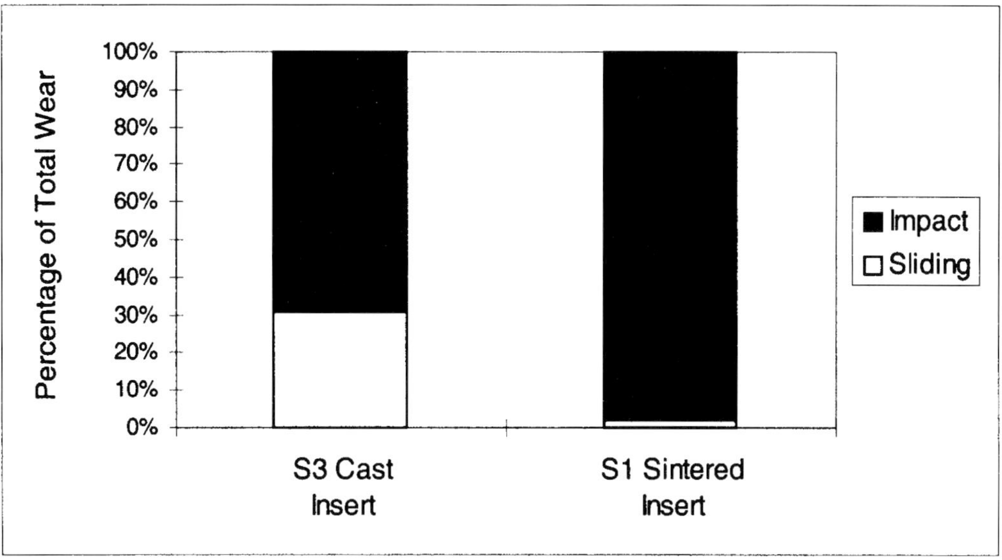

Fig. 7.13 Percentage of total wear caused by impact and sliding for a cast seat insert (S3) engine test

7.2.4.2 Bench tests

Figures 7.14 and 7.15 illustrate recession predictions using the model for tests run on the hydraulic loading apparatus. Initial conditions used for each test simulated are shown in Table 7.3. As can be seen, the model produces a good approximation of valve recession.

Figure 7.16 shows the contributions of impact and sliding wear to the total wear for each case. In general, the effect of impact is reduced, compared to the engine test prediction, as a result of the lower valve closing velocities experienced in the hydraulic loading apparatus. The two exceptions are the lubricated S3 cast seat insert, where the lubricant reduced the sliding wear to about 0.1 per cent of the total, and the oilite seat, which has a very low resistance to impact.

It has been shown that the valve recession model produces good predictions of valve recession for both engine tests and bench tests. It could clearly be used, therefore, to give a quick assessment of the valve recession to be expected with a particular material pair under a particular set of engine/test rig operating conditions. This will help speed up the process undertaken in selecting material combinations or in choosing engine operating parameters to give the least recession with a particular combination.

A quick examination of the model immediately reveals how wear can be reduced by varying engine operating parameters and material properties. For example, reducing valve closing velocity, valve mass, and valve seating face angle, or increasing the valve head stiffness and seat material hardness will reduce valve recession.

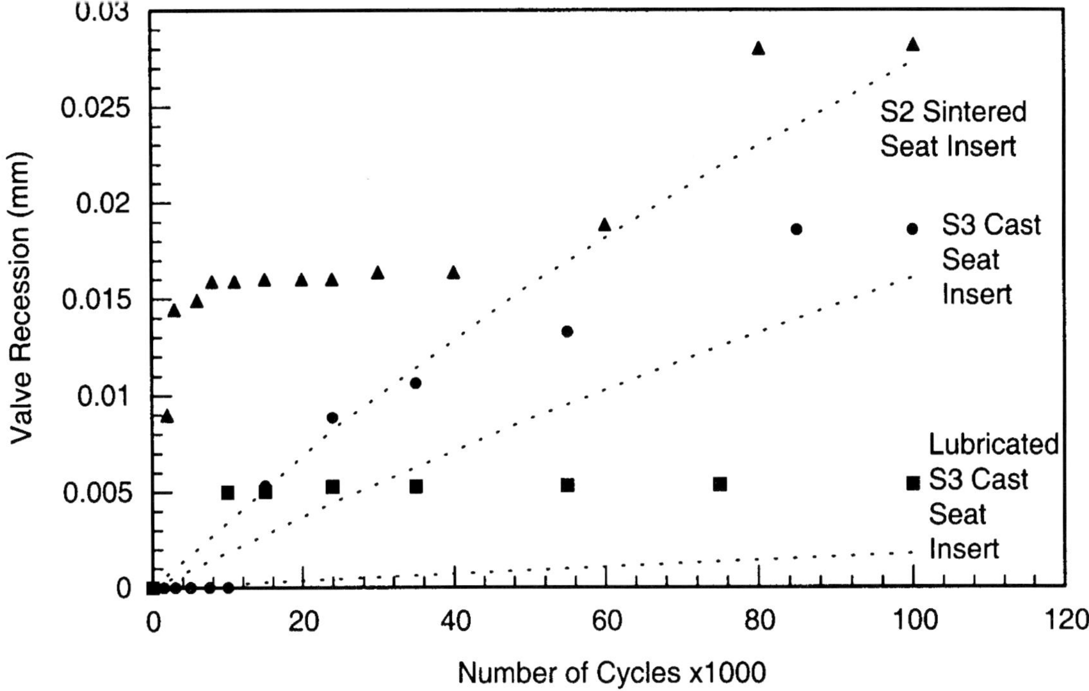

Fig. 7.14 Model prediction (broken lines) versus hydraulic loading apparatus data for a cast seat insert (S3), sintered seat insert (S2), and a lubricated cast seat insert (S3). (Reprinted with permission from SAE paper 2001-01-1987 © 2001 Society of Automotive Engineers, Inc.)

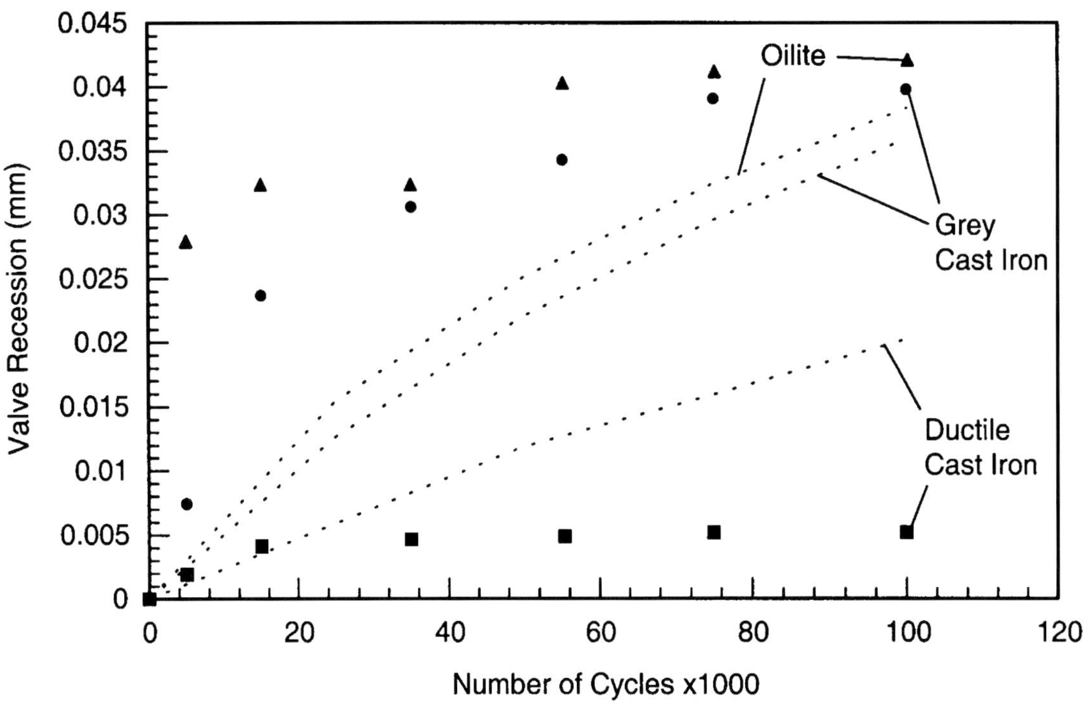

Fig. 7.15 Model prediction (broken line) versus hydraulic loading apparatus data for grey cast iron, ductile cast iron, and oilite seats. (Reprinted with permission from SAE paper 2001-01-1987 © 2001 Society of Automotive Engineers, Inc.)

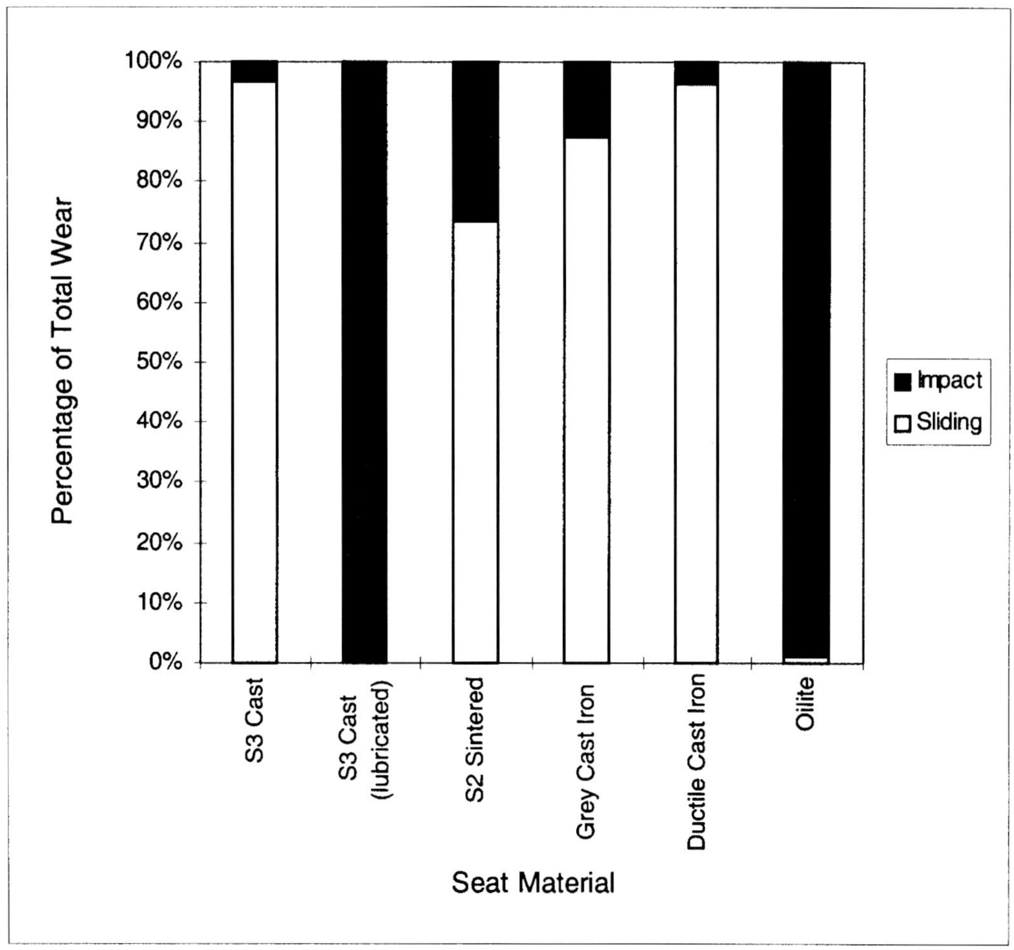

Fig. 7.16 Percentage of total wear caused by impact and sliding for each bench test modelled

By studying the individual contributions of impact and sliding wear it is possible to focus on the particular parameters that need to be altered in order to produce the largest feasible reduction in the total wear for a particular material combination.

In order to improve the model it would be beneficial to be able to relate the impact wear constants K and n to material properties. Currently they have to be determined from test data.

It is likely that as the tests progress the hardness at the seating face of the seats increases. It would also be desirable to include such data in the model.

When selecting wear coefficients using Fig. 7.4 there was a large range of possible values for each material combination/lubrication state being considered (at least one order of magnitude). In order to further improve the model it would be beneficial to obtain more accurate wear coefficient data referring to the actual material pair used.

7.3 Reducing valve recession by design

The flow charts included as Figs 7.17 and 7.18 encapsulate the review of literature, analysis of failed specimens, bench test work, and modelling to create flow charts. The first of these was produced to help reduce the likelihood of valve and seat wear occurring during the design process of a new engine or in the modification of an existing engine (see Fig. 7.17). It was developed directly from the valve recession model and looks at ways in which wear can be reduced by varying the model parameters. The chart looks at sliding and impact wear separately, as they are treated as such in the model.

Changing engine design variables may have an adverse effect on engine performance. For example, a reduction in the peak cylinder pressure could reduce the wear, but could also reduce the engine thermal efficiency if a decreased compression ratio is used. Likewise, stiffening the valve head to reduce the relative movement at the seating face by adding material or altering the valve seating face angle could adversely affect air flow characteristics, and if the valve weight increases significantly, it may affect valve-train dynamics. The introduction of lubrication at the valve/seat interface is clearly unacceptable as this goes against the original stated intention to reduce oil in the air stream of diesel engines. A possible alternative could be the use of solid lubricants incorporated in sintered seat insert materials. However, the problems involved with the use of such materials have been highlighted and some work is needed to improve their performance, certainly with respect to resistance to impact.

Given the problems outlined above and the constant demand for improved performance and efficiency and reduced emissions, reducing the likelihood of valve recession occurring by design may prove impossible. However, wear reduction through the use of new materials, materials selection, or the use of wear-resistant coatings provides an acceptable alternative. As already discussed, the two test rigs developed are suitable for use in the testing of new valve, seat, and valvetrain designs and material ranking.

7.4 Solving valve/seat failure problems

The second flow chart (see Fig. 7.18) was produced to assist in solving valve/seat failure problems. It was developed using experience gained during the progression of the work on failure analysis, experimental work and modelling, and using information obtained in the review of literature.

Three stages to solving a valve/seat failure problem were identified.

i) Determination of valve/seat failure

During engine testing the criterion for determining valve/seat failure is pressure loss in a cylinder.

ii) Analysis of problem

This involves establishing what has happened and why it has happened. Techniques that may be used in this process are:

- profilometry;
- ovality (of seating faces);
- visual inspection;
- optical microscopy.

The profilometry and ovality measurements will show how much wear has occurred and whether it is even. Visual inspection should indicate whether deposit build-up played any part in the failure, and optical microscopy will reveal the wear mechanisms that have occurred which should indicate what has caused the wear to occur (impact or sliding).

iii) Solving the problem

This involves establishing ways in which the causes of the failures identified in stage (ii) can be eliminated. The flow chart goes through the stages detailed above, giving likely reasons why a valve or seat may have failed and possible causes of these failures and then goes on to offer solutions to the problems. This information should help focus any engine testing required to solve a problem, thereby saving time, effort, and cost. This chart is intended to be a working document that should be added to as work is carried out or as different problems arise.

The flow chart can be used to solve problems unique to one cylinder or common to several/all of the cylinders in a head.

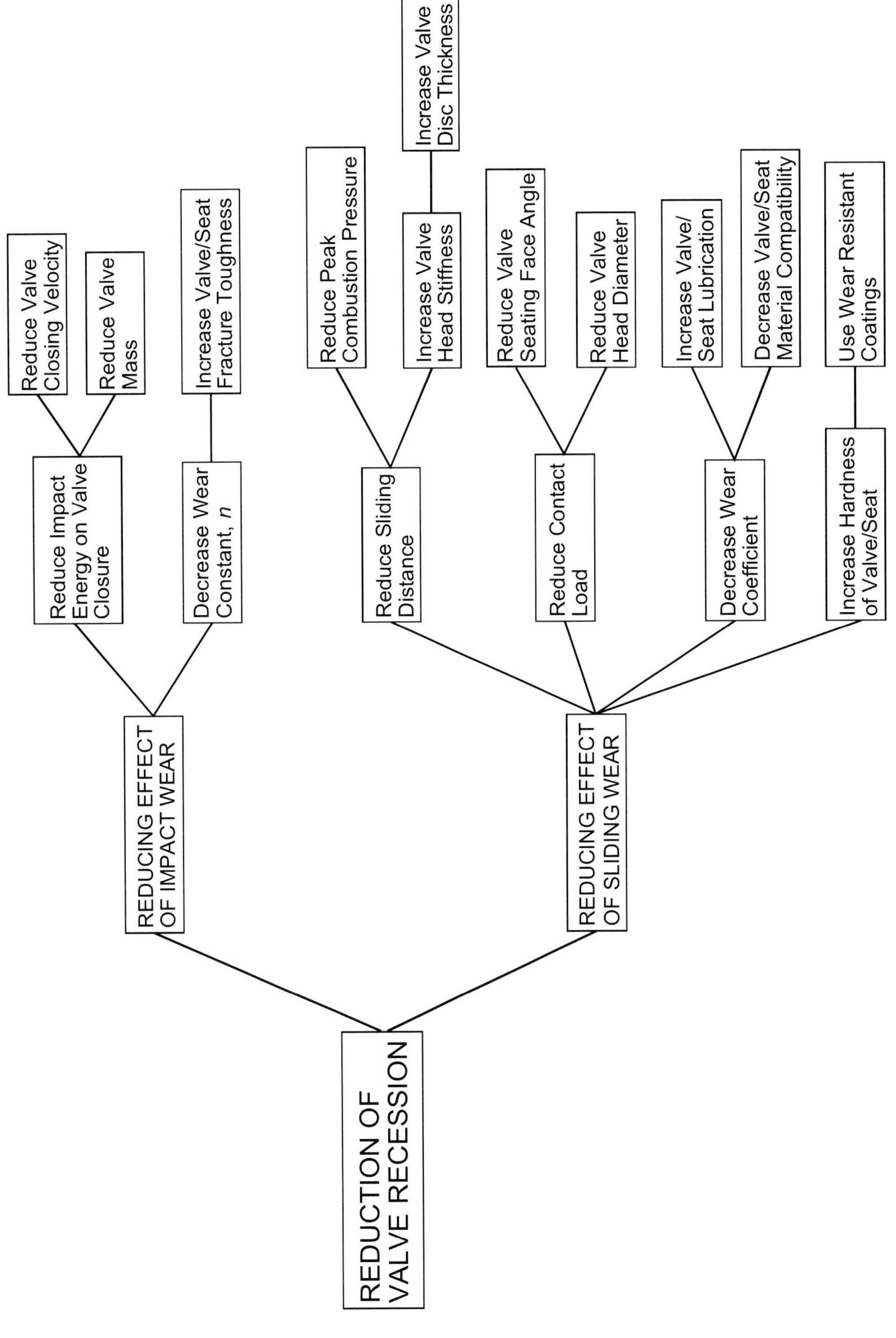

Fig. 7.17 Reducing valve recession by design.
(Reprinted with permission from SAE paper 2001-01-1987 © 2001 Society of Automotive Engineers, Inc.)

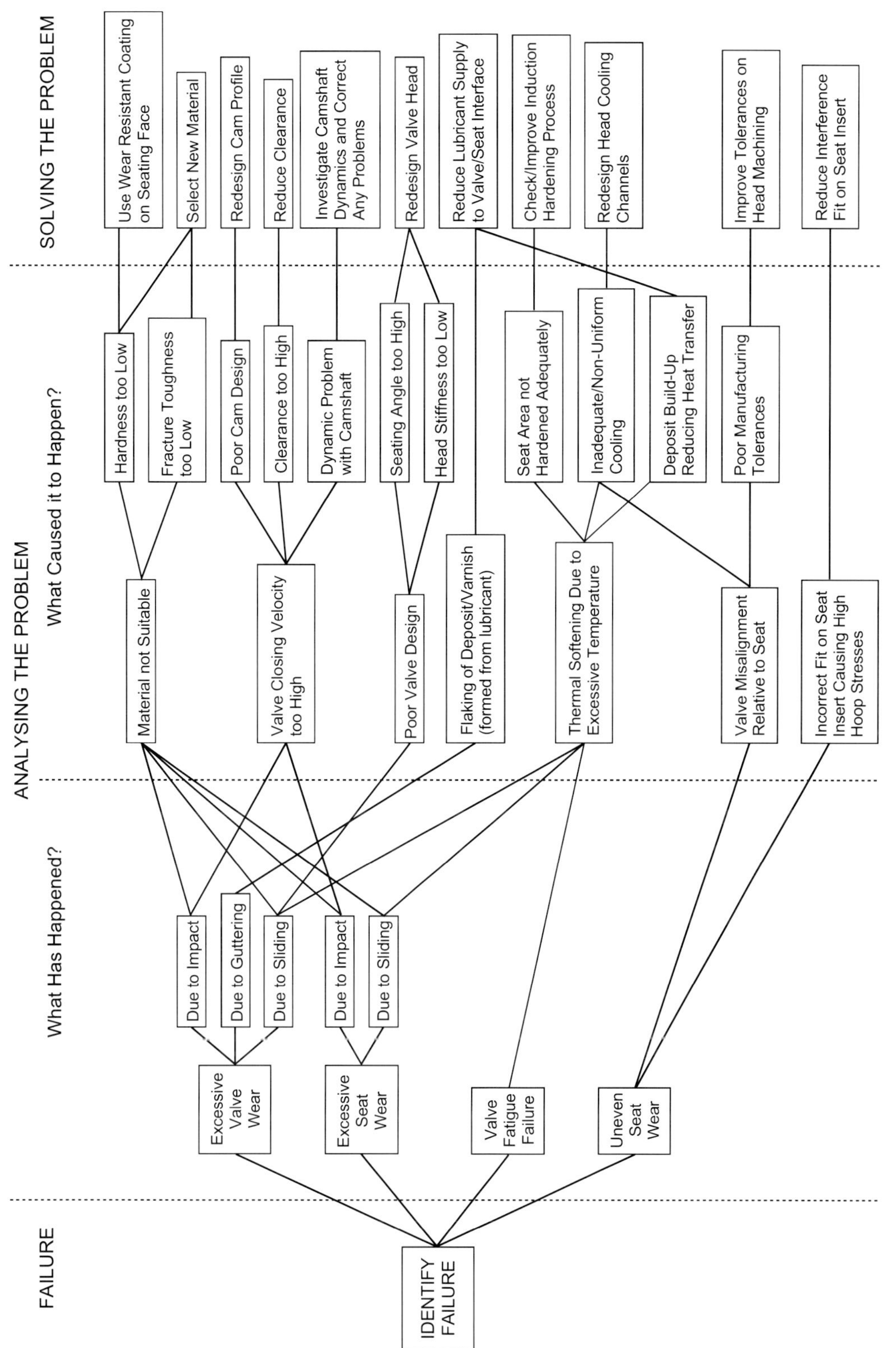

Fig. 7.18 Solving valve/seat failure problems.
(Reprinted with permission from SAE paper number 2001-01-1987 © 2001 Society of Automotive Engineers, Inc)

7.5 References

1. **Pope, J.** (1967) Techniques used in achieving a high specific airflow for high-output medium-speed diesel engines, *Trans. ASME J. Engng Power*, **89**, 265–275.

2. **Giles, W.** (1971) Valve problems with lead free gasoline, SAE Paper 710368.

3. **Archard, J.F.** (1953) Contact and rubbing of flat surfaces, *J. Appl. Physics*, **24**, 981–988.

4. **Suh, N.P.** and **Sridharan, P.** (1975) Relationship between the coefficient of friction and the wear rate of materials, *Wear*, **34**, 291–299.

5. **Stower, I.F.** and **Rabinowicz, E.** (1973) The mechanism of fretting wear, *Trans. ASME, J. Lubrication Technol.*, **95**, 65–70.

6. **Mathis, R.J., Burrahm, R.W., Ariga, S.,** and **Brown, R.D.** (1989) Gas engine durability improvement, Paper GRI-90/0049, Gas Research Institute.

7. **Rabinowicz, E.** (1981) The wear coefficient – magnitude, scatter, uses, ASME Paper 80-C2/LUB-4, *Trans. ASME J. Lubrication Technol.*, **103,** 188–194.

8. **Zum-Gahr, K.H.** (1987) *Microstructure and wear of materials*, Tribology Series No. 10, Elsevier, Amsterdam.

9. **Fricke, R.W.** and **Allen, C.** (1993) Repetitive impact-wear of steels, *Wear*, **163**, 837-847.

10. **Wellinger, K.** and **Breckel, H.** (1969) Kenngrossen und Verscheiss Beim Stoss Metallischer Werkstoffe, *Wear*, **13**, 257–281, in German.

11. **Hutchings, I.M., Winter, R.E.,** and **Field, J.E.** (1976) Solid particle erosion of metals: the removal of surface material by spherical objects, *Proc. R. Soc. Lond. Ser A.*, **348**, 379–392.

12. **Lewis, R.** (2000) *Wear of diesel engine inlet valves and seats*, PhD Thesis, University of Sheffield, UK.

Index